10th International Conference on Mechanical and Manufacturing Engineering (ICME2019)

IOP Conference Series: Materials Science and Engineering
Volume 824

Ayer Keroh, Malaysia
29 - 30 August 2019

Editors:

Zaleha Binti Mohamad
Saliza Azlina Binti Osman
Kamarul Azhan Bin Kamaruddin

ISBN: 978-1-7138-2677-4
ISSN: 1757-8981

Printed from e-media with permission by:

Curran Associates, Inc.
57 Morehouse Lane
Red Hook, NY 12571

Some format issues inherent in the e-media version may also appear in this print version.

Institute of Physics
Dirac House, Temple Back
Bristol BS1 6BE UK

Phone: 44 1 17 929 7481
Fax: 44 1 17 920 0979

techtracking@iop.org

Additional copies of this publication are available from:

Curran Associates, Inc.
57 Morehouse Lane
Red Hook, NY 12571 USA
Phone: 845-758-0400
Fax: 845-758-2633
Email: curran@proceedings.com
Web: www.proceedings.com

TABLE OF CONTENTS

PREFACE

The 10th International Conference on Mechanical and Manufacturing Engineering (ICME 2019) organised by Faculty of Mechanical and Manufacturing Engineering, Universiti Tun Hussein Onn Malaysia is yearly event focusing solely on the Mechanical and Manufacturing in academic and industrial research. The conference was held in Melaka, from August 29 to August 30, 2019.

This annual conference is aimed at highlighting the progress in mechanical, materials and manufacturing engineering and carries the theme “Emerging trends in Academic- Industrial collaboration". ICME 19 aims to provide a platform for academics, students, researchers and industrial players to share knowledge and ideas, widen networking, present their research findings and explore any other potential benefits that might contribute to the better future.

All articles published in this volume IOP Conference Series: Materials Science and Engineering have been peer reviewed through processes administered by the conference organizer. Each article was peer-reviewed by at least two reviewers from the ICME committee and appointed reviewers based on their field of expertise. Reviews were conducted by expert referees to ensure the manuscript is up to the standard. We would like to thank the organizing committee and the members of reviewers for their kind assistance in reviewing the papers.

We would also extend our gratitude for participants in our conference for their invaluable contribution and worthwhile ideas shared in the conference. It hope the proceedings will be helpful to all of you.

The Editors

Dr. Zaleha Binti Mohamad

Ir. Dr. Saliza Azlina Binti Osman

Dr. Kamarul Azhar Bin Kamaruddin

IOP Conf. Series: Materials Science and Engineering **824** (2020) 011002 doi:10.1088/1757-899X/824/1/011002

Peer review statement

All papers published in this volume of *IOP Conference Series: Materials Science and Engineering* have been peer reviewed through processes administered by the proceedings Editors. Reviews were conducted by expert referees to the professional and scientific standards expected of a proceedings journal published by IOP Publishing.

IOP Conf. Series: Materials Science and Engineering **824** (2020) 012001 doi:10.1088/1757-899X/824/1/012001

Influences of Starch on Ceramic-Foam Fabrication: A Short Review

Abdul Abbas N.H Yatim[1,2,a] **and H.A. Rahman**[1,b*]

[1] Faculty of Mechanical and Manufacturing Engineering,
Universiti Tun Hussein Onn Malaysia,
86400 Parit Raja, Batu Pahat, Johor, Malaysia

[2]Politeknik Ungku Omar Jalan Raja Musa Mahadi,
31400 Ipoh, Perak, Malaysia

E-mail: [a]nhamidah@puo.edu.my, *[b]hamimah@uthm.edu.my

Abstract. Ceramic foams are high-value products with multiple functions, such as thermal insulation, acoustic insulation, fireproof and freezing tolerance. Wide-ranging ceramic-foam applications can reduce energy consumption and indoor fire hazards. The utilisation of ceramic materials in manufacturing requires appropriate methods to reduce the problems associated with these features. Ceramic foams have excellent thermal stability and structural resilience against organic solvents and chemical corrosion. They can also be used in various engineering applications. Natural additives such as starch, sucrose and yolk are used to control the pore structure of the products. These additives serve as foaming and binding agents to enhance the stability of foaming slurry. The use of starch strengthening in the ceramic fabrication is widespread because of its low cost and reduced impact on the environment. The effects of starch addition on the mechanical properties of ceramic foams include increased density, linear shrinkage and compressive strength, as well as decreased porosity and water absorption. These effects depend on the contents of starch and sintering temperature. This review focuses on the use of starch as a binding agent for ceramic-foam fabrication.

1. Introduction

Ceramic foams are inorganic materials formed from a three-dimensional ceramic fibre network. Ceramic foams have low density, high specific area, controlled permeability, thermal insulation and good acoustic as well as very strong chemical resistance. Ceramic foams with open pore structure are mostly used as carriers for catalysts, filters, bioreactors and bone substitutes, while ceramic foams with closed pore structure are used for sonic insulation and lightweight structured components. Ceramic foams are a specialised class in porous ceramics and have high structural and porosity. The spherical pores in ceramic foams are randomly arranged and closely stack together [1]. Ceramic foams with special combination properties, such as thermal conductivity and water absorption, have attracted the attention of manufacture using it as heat insulation, sound absorption and shock wave-absorbing construction [2]. Special characteristics of ceramic foams, such as great surface area, high porosity, low density, high refractoriness, good chemical resistance, long severe environment and low thermal

IOP Conf. Series: Materials Science and Engineering **824** (2020) 012001 doi:10.1088/1757-899X/824/1/012001

conductivity, have attracted several fields for use. These features make ceramic foams be widely used in many respects, including thermal insulation, fire resistance, waterproof and chemical stability. With the excellent combination of their special characteristics, ceramic foams have been widely used in the area of construction insulation [3].

Ceramic foams have been widely utilised because of their usability in some areas of technology. Among other things, ceramic materials are used as insulation refractories, catalyst support and filters for molten metals [4]. Ceramic foams are also increasingly being used in various engineering fields as catalyst, liquid or gas filtrate, extractor, separator, thermal insulator, scaffolding for biological and special low application and gravity components for structural applications. Ceramic foams that have such characteristics as high permeability, relatively low density, high surface area and low thermal conductivity, are therefore indispensable nowadays. The usage of ceramic foams depends on the technique used in their fabrication [5].

Various methods, such as particle stacking, polymeric sponge technique, sacrificial template technique, freezing casting and direct foaming technique, are used to fabricate ceramic foams [4]. Another technique to fabricate ceramic foams such as particle organisers, replication of sacrificed template, burning out of fugitive pore formers and direct foaming of colloidal slurry [6]. Moreover, several techniques, including replicas, templating techniques and direct foaming, have been developed to produce ceramic foams [7]. Ceramic foams are ecologically porous materials with excellent characteristics, such as low density, high specific strength, good thermal insulation and chemical constancy. Current techniques used to produce ceramic foams include sol-gel, sacrificed template, addition of pore-forming agent and foaming technique of powder [8]. Foams can be defined as a spreading system based on the bubble gas of their standard phase, air, dispersed in a continuous liquid or solid matrices (solid foam), such as steel, glass and concrete foam. Drenckhan, W.,et al (2010) studied foam preparation to evaluate the stability of foaming time and the control of bubble size [9]. The structure, coarsening process, drainage and dynamic properties of foams are influenced by fine particles [10]. Various methods have been established to acquire the best ceramic foams for different engineering applications. Each of the techniques used to produce ceramic foams, however, has respective weaknesses. Among the methods commonly used in fabricating ceramic foams are gel casting, freeze casting, extrusion, compaction, starch consolidation casting (SCC) sol–gel synthesis, organic foaming method and using pore-forming agent.

2. Types of methods for ceramic-foam fabrication

Many methods have been developed to fabricate ceramic foams. The preparation of ceramic foams has significantly affected the microstructure/ceramic-foam properties produced. Ceramic foams delivered by replica techniques, sacrificial templates and direct bubbling methods typically have a high porosity [11]. Ceramic foams can be formed through different methods, and the most conservative technique is the reticulated sponge method or replica technique. By this method of replication, the polymer span is coated with a ceramic slurry, dried, covered and sintered [12]. Ceramic foams have been widely used as thermal insulator, catalyst support, filter and medical implants because they affect the cellular microstructure of the ceramic foams. Several ceramic foam-processing methods, such as replica and direct foaming methods, have been developed. The direct foaming technique is an interesting technique, in which foams are made with open cells or closed cells [13]. Ceramic foams have multipurpose pore density, surface range and chemistry [14]. The actual cell shape, however, depends on some factors, including the manufacture process. Ceramic foams can also be produced via starch consolidation, uniaxial pressing, polymeric sponge replication method and foaming method.

Porous ceramics have been produced by many processing methods, such as sol–gel synthesis, freeze casting technique, organic foaming technique and pore-forming agent technique. Among the methods, the pore-forming agent method is most usually used and is an effective method; pore-forming agents, such as starch, graphite, organic particulates or saw dust, have been used usually. Nevertheless, most of these pore-forming agents only form structures with casually organised irregular pores with several sizes in the final structure caused by agglomeration and nonuniformity of pore-forming agents in raw materials; such condition is threatening to be avoided [15]. Different methods, such as gel casting, freeze casting, extrusion, compaction and SCC, have been used to fabricate cellular ceramic techniques. In the fabrication of porous ceramics, various types of binders, such as potatoes, corn, potato, rice, wheat and poppy seeds, can be used. The use of starch take out from seeds of cereals (corn, sorghum, wheat or rice), certain roots (potato, cassava or tapioca or arrowroot) and the sago palm has been applied in the use of commercial ceramics. Starch, as an additive, is a material that has the features of easy processing, commercial availability and burning completely without residual ash. Sago starch has great potential as a binder and a pore-forming agent in the manufacture of cellular ceramics [16]. The main benefit of using starch as a consolidation agent for ceramic slurries compared with other biopolymers is that it is a liquid and secures the characteristic of being residual-free through starch burnout. This method is a technology that is eco-friendly and environmentally friendly to be implemented in the processing of ceramic materials with various applications [17].

3. Starch consolidation

The starch consolidation method for ceramic-foam characteristic produces the capability to get a porous body with complex shapes and controlled microstructures by using early suspensions with dissimilar solid loading and can be formed in multiple applications. The use of starch from various sources provides the resulting ceramic body with different porous morphologies. The most commonly used starch examples are rice, potatoes and corn. The potato starch shows the biggest and most anisometric granules, whereas rice starch is the least and more isometric. A technique that uses starch as a consolidation agent for tricapital pore-structured silica bodies adopts two consolidation methods, starch and foaming [18]. The starch consolidation way for ceramic slurries, as one of the consolidation procedures, has been usually explored. In this process, starch is loaded in the aqueous slurry of the ceramic powders. Direct consolidation is a novel process based on colloidal processes. This process has the potential to form complicated ceramic bodies from minor to huge components with exact porosity, defects and density. The powder used in this technique should be fully smoothed to prevent the decline in the strength and reliability of the ceramic body produced [17].

SCC techniques are based on the capability of starch to swell and ultimately gelatinise in the water at a temperature of 60 °C–80 °C. Ceramic green samples can be formed from suspension in resistant moulds (usually metal moulds). When the granules or starch particles swell, they will also entertainment as binders adding strength to the unified body, which allows it to form before drying. After burning out of the starch and sintering of the ceramic matrix, a material is gained with a porosity equivalent to the original amount, shape and size of the starch particles [19]. Starch consolidation techniques have been broadly explored to form ingredients for ceramics through the use of starch loaded in ceramic slurry. The inactivity of the starch in the ceramic slurry enables the formation of the ceramic green body through slip casting method into a nonporous mould [20]. This condition is due to the casting of starch strengthening categorized as a direct pouring method with no drainage required to form the desired ceramic.

Sopyan, I., and Fadli, A. (2012) developed a method for protein foaming and consolidation by using yolk as a pore-forming agent to fabricate porous ceramics using alumina as raw materials [21]. Alumina powder and egg yolks were stirred in a beaker with ratio of alumina of 0.65 (slurry A), 0.75 (slurry B) and 1.00 (slurry C). Magnetic powder was blended at 150 rpm for 3 h at room temperature. The slurry was put into a cylindrical open stainless steel mould and heated in an oven at 180 °C for 1 h. Castor oil was used as a lubricant to facilitate mould release. The samples were sintered in a furnace at 10 °C/min up to 600 °C to remove egg yolks and then at 2 °C/min up to 1550 °C.

The diagram of the processing scheme for porous silica ceramics using foaming and starch consolidation technique is presented in Figure 1. A slurry is ready with silica powder of 0.2wt% CMC and 10wt% of cassava starch and added with distilled water by a planetary ball mill for 30 min. Lactic acid with pH value of 3–4 was added to the mixture to complete the slurry. After consolidation and cooling, the mould is left at room temperature for 36 h and then stored at 50 °C–80 °C for 2–3 days.

Complete drying is conducted at 120 °C, and burnout is completed by heating in air at 1.0 °C min^{-1} to 500 °C, with holding time of 2 h at 300 °C to avoid violent contamination. The mould is then sintered at 1250 °C for 5 h [17].

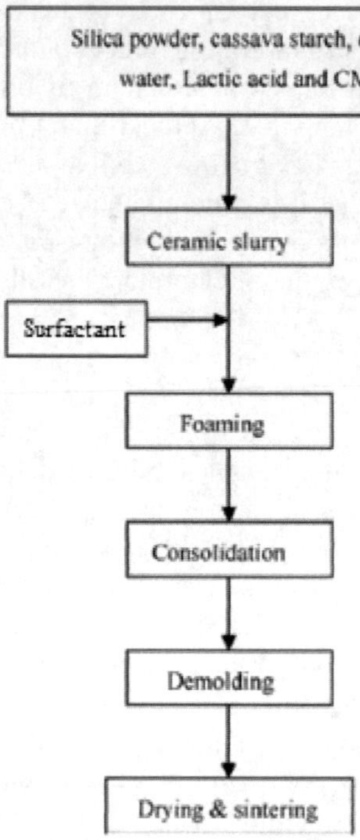

Figure 1. Flow chart of the process for porous silica ceramics using foaming and starch consolidation method [22].

Mao, X, et al. (2008) stated that the foaming and starch combination technique is suitable for porous ceramic fabrication [22]. Techniques that combine foaming and starch consolidation can produce extremely porous ceramics and are also economical, practical and environmentally friendly [23]. The foaming and starch consolidation method includes ceramic suspensions, foaming, consolidation, drying and sintering, as described in Figure 2.

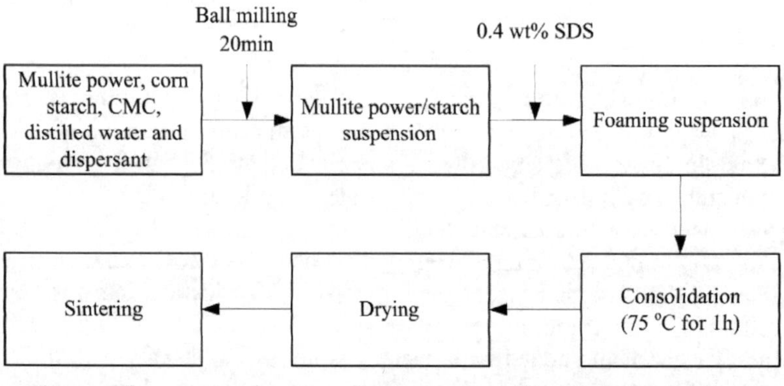

Figure 2. Flow chart of the process for preparing porous mullite ceramics by foaming and starch consolidation [23].

foam fabrication. This process uses the capability of starch to form pores and granule binders in ceramic foams. The starch can swell and calibrate at a certain temperature of 60 °C–80 °C and a sintering temperature of 1250 °C. This swelling is due to the amount of water absorbed by starch. Swelling improves the strength of the consolidated body. This technique can control the size of pores and the resulting pore distribution by altering the amount of starch used and the type of starch used. Starch is used as a pore-forming agent because of the naturally occurring chemical properties that can guarantee the absence of waste after combustion.

4. Types of starch

Lyckfeldt, O, et al. (1998) stated that starch grains are usually white, dense and unsolvable in water at room temperature and comprise a mixture of two polysaccharides, a linear amylose and a split amylopectin [24]. Amylose gives its soil starch in aqueous suspension. The polymer chains in the starch are formed by glucose units that produce large amounts of hydrophilics which provide strength to starch granules, as shown in Figure 3. This condition produces good features, such as coagulating, stabilising, membrane-forming and gelling properties [19].

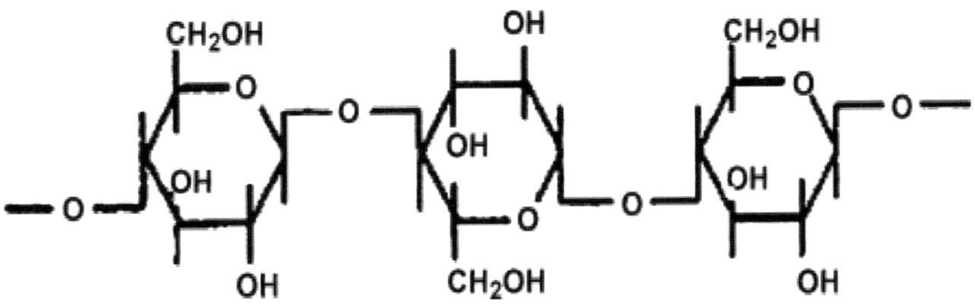

Figure 3. Starch is a polymer consisting of condensed glucose units [19]

Binders are a long chain of polymer networks that work to provide strength to the green body by building bridges among particles. In the drying process, binders are removed, leaving the holes in the frame or cell wall. The traditional binders are in the form of starches obtained from plants, such as tubers or roots and cereals. They are a major source of eco-friendly materials that are easily processed and operated, low cost, consistent and have good and biodegradable distribution of shapes and sizes. Existing biopolymers consist of condensed glucose units. The dissolution of part of glucose starch molecules causes them to undergo irreversible swelling and crystalline crystallisation called starch gelatinisation and alter from pure suspension to elastic gel [16]. Gregorova, E, *et al.* (2010) demonstrated that various types of binders, such as potatoes, corn, cassava, rice, wheat and poppy seeds, are used for porcelain ceramics [25]. No literature related to the use of sago (*Metroxylon sagu*) as a starch source and a binder to produce ceramic cells is available. Sago has the same properties as those of binders; it is available in Malaysia at a low cost and can be used as a component used in adhesive production. Sago is gelatinised because it has a low gelatinisation temperature, has high viscosity and is easily formed because of low-syneresis gel [16].

Starch is a high molecular carbohydrate comprising linear amylose and amylopectin [17]. When in the water at a high temperature, starch experiences glutamine, followed by hydrolysis containing amylose. Amylose and amylopectin are also known as polysaccharides with much sugar and are formed from the combination of several glucose units that form long chains. They consist of two types of macromolecules having different glucose monomers. Glucose units are arranged in an alternate arrangement. Amylose is largely linear polymer having a repeating hydroxide bond. Starch can increase viscosity, especially at low shear rates. When the shear rate is high, the polymer leads to the

IOP Conf. Series: Materials Science and Engineering **824** (2020) 012001 doi:10.1088/1757-899X/824/1/012001

flow of visible clear viscosity. The phenomenon is due to the amylopectin effect, which spreads into its structural solution, such as the tree, thereby keeping the particles in the distance to escape separation, as shown in Figure 4. Changes affect yield stress at time out but do not show a strong impact on plastic viscosity. The weight of starch molecules may show a role in the stabilisation influence, and the ratio between amylose and amylopectin may be very high [1].

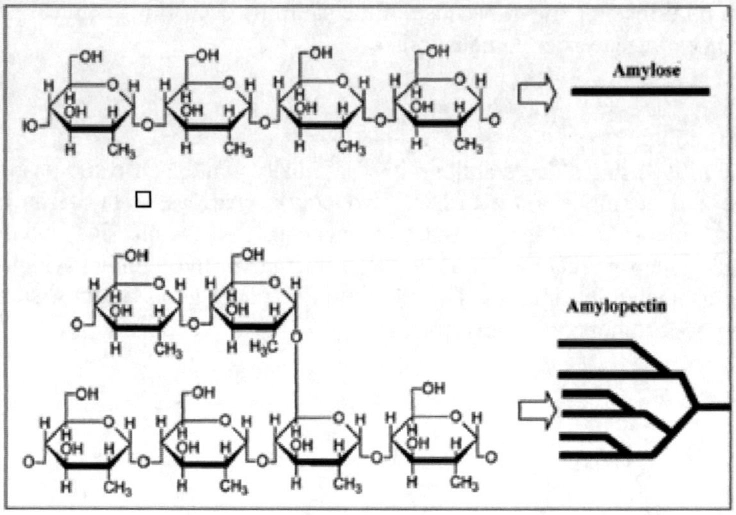

Figure 4. Chemical structure and performance of starch components in aqueous media: (a) chemical structure of amylose and amylopectin [17].

Starch is from several types of cereals, such as corn, rice, flour, potato and cassava. Starch is widely used in manufacturing process as a component in the field of ceramic processing and as an additive in the textile industry, paper manufacture, food processing, pharmaceuticals and building materials. Among the characteristics of starch that allow them to be widely used are the increase in viscosity, gelatin, fasteners and the ability to form. Table 1 and Figure 5 show the type and amount of the starch used in the various fabrications of ceramic foams from previous studies.

Table 1. Types and amount of starch used in the fabrication of ceramic foams.

No.	Types of starch	Amount	Raw material	Reference
1.	Wheat starch, Potato and corn	10–50vol%	Alumina powder	[6]
2.	Cassava	10–20vol%	Alumina oxide	[2]
3.	Corn	3–13mass%	Calcined alumina	[13]
4.	Potato	10–40vol%	HA	[1]
5.	Sago and Tapioca	50–70wt% (solid weight)	Porcelain raw powder	[12]
6.	Cassava	25wt%	Sayong ball clay	[24]
7.	Corn	10wt%		[5]
8.	Cassava	10wt%	Silica	[17]
9.	Wheat flour	4 g	HA	[19]

IOP Conf. Series: Materials Science and Engineering **824** (2020) 012001 doi:10.1088/1757-899X/824/1/012001

5.Influences of starch on the mechanical properties of ceramic foams

After the formation of ceramic foams through the drying process, several mechanical, physical and chemical tests are performed to study the characteristics of ceramic-foam fractions. Mechanical tests are conducted to evaluate the compressive strength of ceramics formed. For features, linear shrinkage, foam morphology, porosity and density are studied through physical testing. Mechanical properties are particularly emphasised in the characterisation of ceramic-foam structures. The compressive strength of ceramic foams is evaluated to test the strength of ceramic foams, which are designed to have a high porosity of more than 80%. Compressive strength affects the porosity, size, distribution and morphology of pores. The porosity of a ceramic foam utilizes a high effect on compressive strength, because it determines the solid cross-sectional load bearing of ceramic foams [6].

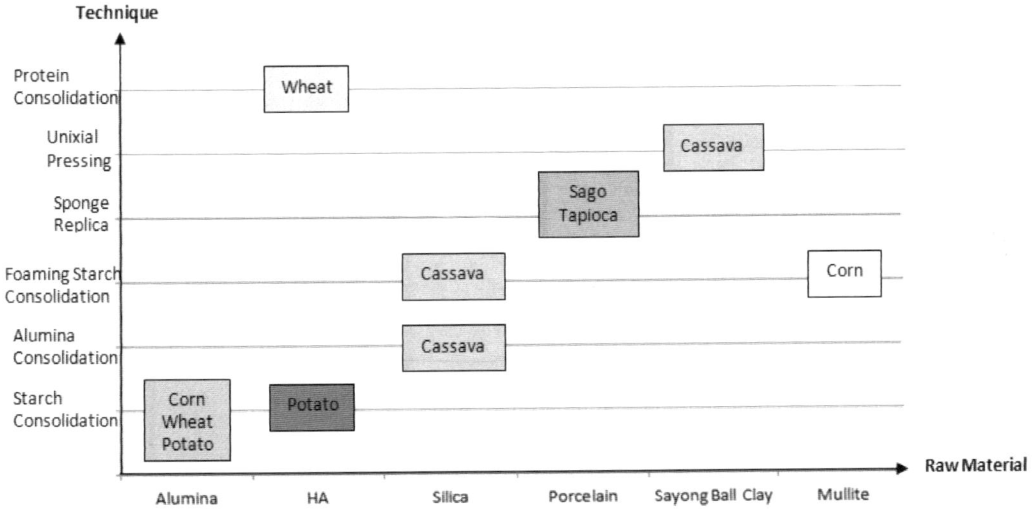

Figure 5. Summary of starch used.

5.1 Density and porosity

Starch content effects the apparent porosity and bulk density of porous alumina samples [19]. Archimedes water displacement procedure is used to measure the bulk density and total porosity of sintered samples [16]. Bulk density and open porosity are determined by the Archimedes method in water [20]. Table 2 presents that previous studies showed that linear shrinkage will change according to the starch composition used.

IOP Conf. Series: Materials Science and Engineering **824** (2020) 012001 doi:10.1088/1757-899X/824/1/012001

Table 2. Density and porosity of porous ceramic body.

No.	Technique of fabrication	Starch amount	Linear shrinkage (%)	Density, g/cm^3	Porosity, %	Reference
1	Starch consolidation casting	10–50vol%		-	20–50	[6]
2	Starch consolidation casting	10–40vol%		1.36–1.65	79.2–90.2	[1]
3	Starch consolidation casting	3–13mass%	0.00%-9.44%	1.4–2.05	46–64	[13]
4	Alumina consolidation	10–20vol%	15.2%-29.6%	0.4–0.75	13–15	[2]
5	Foaming and starch consolidation	10wt%		18–30	-	[17]
6	Sponge replica	50–70wt% (solid weight)		0.37–0.51	79–81	[12]
7	Uniaxial pressing	25wt%		-	-	[24]
8	Protein foaming consolidation	4 g	26–77vol%	-	46–52	[19]
9	Foaming and starch consolidation	10wt%		-	86–73	[5]

5.2 Shrinkage

Linear shrinkage occurs due to a defined sintering process as the decline or shrinkage of a percentage of the length of a ceramic sample, and it is important in the ceramic industry. Linear shrinkage tests are performed to examine the sample shrinkage rate produced by the drying process. Length and dimensions are required in several applications to determine the amount of ceramic samples produced before the formation process. Linear contraction is due to changes in particle size and sample size. Linear shrinkage is almost inversely proportional to the radius of particles.

Shrinkage is measured using a slide calliper and calculated from mould dimensions and body dimensions after firing [20]. Linear shrinkage depends on the content of starch used and the sintering temperature. The linear shrinkage of porous alumina samples sintered at different firing temperatures of 1400 °C–1700 °C is depicted in Figure 6.

IOP Conf. Series: Materials Science and Engineering **824** (2020) 012001 doi:10.1088/1757-899X/824/1/012001

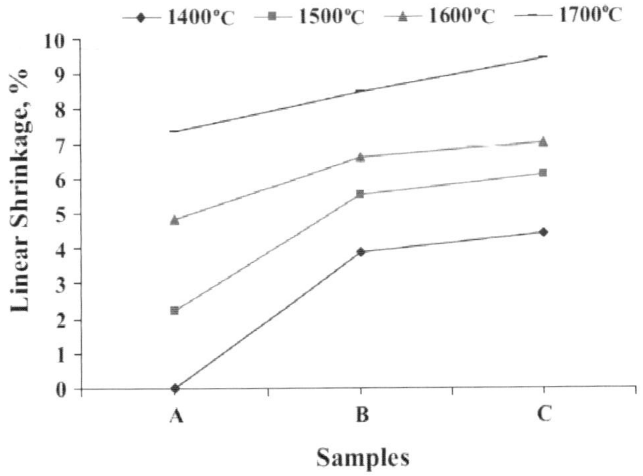

Figure 6. Linear shrinkage of porous alumina samples sintered at different temperatures of 1400 °C, 1500 °C, 1600 °C and 1700 °C [13].

5.3 Compressive strength

The compressive strength test is executed to evaluate the mechanical properties of formed sintered samples using a universal testing machine. The compressive strength is calculated from the maximum point of the stress/strain graph, which occurs when the first crack performed in the samples. The average of three samples is taken as the measure of the compressive strength of the samples [17]. Gong, L, *et al.* (2014) studied the compressive strength of porous mullite ceramics sintered at different temperatures, as shown in Figure 7. For 62.5wt% solid loading, the compressive strength of porous mullite ceramics increases from 1.02 MPa to 13.61 MPa with increased sintering temperature; for 67.5wt% solid loading, it increases from 2.14 MPa to 22.38 MPa. Two factors could impact the mechanical strength of porous mullite ceramics prepared in this work: porosity and sintering neck. From 1200 °C to 1500 °C, the compressive strength increases slightly because the effect of porosity is dominant. After sintering at 1600 °C for 2 h, the samples display a affected improvement in compressive strength. The result is attributed to that the increased sintering temperature leads to improved densification and hence increasingly strong sintering neck, resulting in dramatically increased compressive strength [23]. Table 3 shows the obtained compressive strength in previous studies using different raw materials.

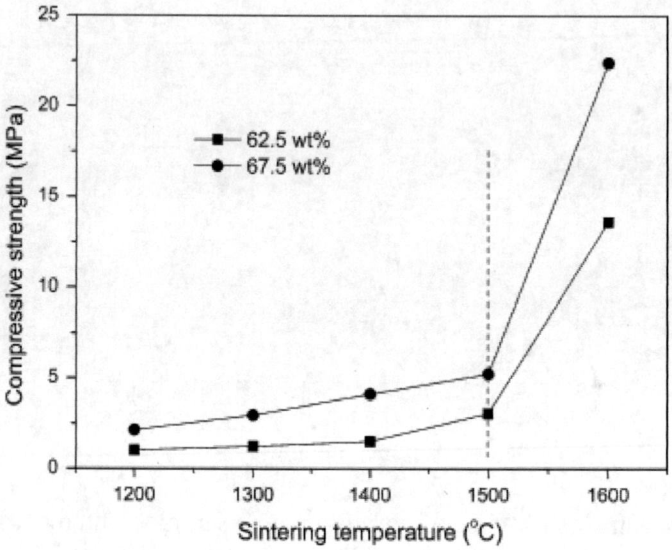

Figure 7. Compressive strength of porous mullite ceramics sintered at different temperatures [23].

On the basis of previous studies using the same technique to form foaming ceramics, the compressive strength of foaming ceramics is closely related to the amount of starch used, material porosity, material density and solid content. Compressive strength increases with decreased material porosity decreases, and material density increases with increased solid content.

Table 3. Compressive strength of porous ceramic body.

No.	Raw material	Compressive strength, MPa	Reference
1	Porcelain raw powder	0.33–0.93	Jamaludin A.R, *et al.* (2014)
2	Porcelain raw powder	0.22–0.58	Jamaludin A.R, *et al.* (2015)
3	HA	5.6–21.4	Ahmed, Y. M. Z, *et al.* (2015)
4	Silica	4.17	Mao, X, *et al.* (2008)
5	HA	0.1–6.4	Sopyan, I., *et al.* (2012)
6	Mullite powder	1–22	Gong, L, *et al.* 2014

6. Conclusion

This study reviews the commonly used starch in the market for ceramic-foam fabrication, especially in starch consolidation techniques. From previous studies, many types of starch, such as corn, cassava, tapioca, potato, wheat flour and sago, are used in ceramic-foam fabrication. The main benefit of using starch as a consolidation agent over other biopolymers is that it is residual-free during burnout. The value of starch content affects the characteristics of fabricated ceramic foams. Ceramic foams containing the same starch content indicate increased density with increased sintering temperature, whereas porosity and water absorption decrease. The values of linear shrinkage and compressive strength also increase with increased starch consumption. This review shows that the sintering temperature used in previous studies ranged from 1200 °C to 1700 °C. Starch is able to swell and gelatinise at temperatures of 60 °C– to 80 °C. Current swelling gives strength to the ceramic-foam body. Starch is used in ceramic-foam fabrication because of its natural properties that guarantee the environment.

Acknowledgement

The authors would like to thank the Ministry of Education Malaysia for supporting this research under Fundamental Research Grant Scheme Vot No. FRGS/1/2016/TK05/UTHM/02/3. This research is partially sponsored by Universiti Tun Hussein Onn Malaysia.

References

[1] Yin, L., Zhou, X., Yu, J., & Wang, H. (2017). Effect of rotating speed during foaming procedure on the pore size distribution and property of silicon nitride foam prepared by using protein foaming method. *Ceramics International*, *43*(5), 4096 4101.

[2] Jiang, C., Huang, S., Li, G., Zhang, X., & Cheng, X. (2018). Formation of closed-pore foam ceramic from granite scraps. *Ceramics International*, *44*(3), 3469-3471.

[3] Ge, X., Zhou, M., Wang, H., Liu, Z., Wu, H., & Chen, X. (2018). Preparation and characterization of ceramic foams from chromium slag and coal bottom ash. *Ceramics International*, *44*(10), 11888-11891.

[4] Hou, L. J., Liu, T. Y., & Lu, A. X. (2017). Red mud and fly ash-based ceramic foams using starch and manganese dioxide as foaming agent. *Transactions of Nonferrous Metals Society of China*, *27*(3), 591-598.

[5] Colonetti, V. C., Sanches, M. F., de Souza, V. C., Fernandes, C. P., Hotza, D., & Quadri, M. G. N. (2018). Cellular ceramics obtained by a combination of direct foaming of soybean oil emulsified alumina suspensions with gel consolidation using gelatin. *Ceramics International*, *44*(2), 2436-2445.

[6] Huo, W., Zhang, X., Chen, Y., Wang, D., Liu, J., Yan, S., & Yang, J. (2018). Mechanical strength of highly porous ceramic foams with thin and lamellate cell wall from particle-stabilized foams. *Ceramics International*. *44*(5). 5780-5784.

[7] Celani, A., Blackburn, S., Simmons, M. J., Holt, L. M., & Stitt, E. H. (2018). Formulation of Ceramic Foams: a New Class of Amphiphiles. *Colloids and Surfaces A: Physicochemical and Engineering Aspects*, *536*, 104-112.

[8] Zhou, M., Ge, X., Wang, H., Chen, L., & Chen, X. (2017). Effect of the CaO content and decomposition of calcium-containing minerals on properties and microstructure of ceramic foams from fly ash. *Ceramics International*, *43*(12), 9451-9457.

[9] Drenckhan, W., & Langevin, D. (2010). Monodisperse foams in one to three dimensions. *Current opinion in colloid & interface science*, *15*(5), 341-358.

[10] Huo, W., Zhang, X., Chen, Y., Lu, Y., Liu, J., Yan, S., & Yang, J. (2017). Novel mullite ceramic foams with high porosity and strength using only fly ash hollow spheres as raw material. *Journal of the European Ceramic Society*.

[11] Li, F., Kang, Z., Huang, X., Wang, X. G., & Zhang, G. J. (2014). Preparation of zirconium carbide foam by direct foaming method. *Journal of the European Ceramic Society*, *34*(15), 3513-3520.

[12] Jamaludin, A. R., Kasim, S. R., Abdullah, M. Z., & Ahmad, Z. A. (2014). Sago Starch As Binder And Pore-Forming Agent For The Fabrication Of Porcelain Foam. *Ceramics International*, *40*(3), 4777-4784.

[13] Khattab, R. M., Wahsh, M. M. S., & Khalil, N. M. (2012). Preparation and characterization of porous alumina ceramics through starch consolidation casting technique. *Ceramics International*, *38*(6), 4723-4728.

[14] Ho, W. (2015). Efficient Photocatalytic Degradation Of NO By Ceramic Foam Air Filters Coated With Mesoporous Tio 2 Thin Films. *Chinese Journal of Catalysis*, *36*(12), 2109-2118.

[15] Xu, G., Ma, Y., Cui, H., Ruan, G., Zhang, Z., & Zhao, H. (2014). Preparation Of Porous Mullite–Corundum Ceramics With Controlled Pore Size Using Bioactive Yeast As Pore-Forming Agent. *Materials Letters*, *116*, 349-352.

[16] Lyckfeldt, O., & Ferreira, J. M. F. (1998). Processing Of Porous Ceramics By 'Starch Consolidation'. *Journal of the European Ceramic Society*, *18*(2), 131-140.

[17] Ahmed, Y. M. Z., Ewais, E. M. M., & El-Sheikh, S. M. (2015). Potato Starch Consolidation Of Aqueous HA Suspension. *Journal of Asian Ceramic Societies*, *3*(1), 108-115.

[18] Almeida, F. A., Botelho, E. C., Melo, F. C. L., Campos, T. M. B., & Thim, G. P. (2009). Influence of cassava starch content and sintering temperature on the alumina consolidation technique. *Journal of the European Ceramic Society*, *29*(9), 1587-1594.

[19] Khattab, R. M., Wahsh, M. M. S., & Khalil, N. M. (2012). Preparation and characterization of porous alumina ceramics through starch consolidation casting technique. *Ceramics International*, *38*(6), 4723-4728.

[20] Gregorova, E., & Pabst, W. (2007). Porosity and pore size control in starch consolidation casting of oxide ceramics—achievements and problems. *Journal of the European Ceramic Society*, *27*(2-3), 669-672.

[21] Sopyan, I., & Fadli, A. (2012). Floating porous alumina from protein foaming consolidation technique for cell culture application. *Ceramics International*, *38*(6), 5287-5291.

[22] Mao, X., Wang, S., & Shimai, S. (2008). Porous Ceramics With Tri-Modal Pores Prepared By Foaming And Starch Consolidation. *Ceramics International*, *34*(1), 107-112.

[23] Gong, L., Wang, Y., Cheng, X., Zhang, R., & Zhang, H. (2014). Porous mullite ceramics with low thermal conductivity prepared by foaming and starch consolidation. *Journal of Porous Materials*, *21*(1), 15-21.

[24] Luyten, J., Mullens, S., Cooymans, J., De Wilde, A. M., Thijs, I., & Kemps, R. (2009). Different methods to synthesize ceramic foams. *Journal of the European Ceramic Society*, *29*(5), 829-832.

[25] Gregorova, E., Pabst, W., Zivcova, Z., Sedlarova, I., & Holikova, S. (2010). Porous alumina ceramics prepared with wheat flour. *Journal of the European ceramic Society*, *30*(14), 2871-2880.

IOP Conf. Series: Materials Science and Engineering **824** (2020) 012002 doi:10.1088/1757-899X/824/1/012002

Potential Energy Saving by ACMV Temperature Setting in University Building

Nurzhatul Aziemah Abdul Omar[1], Noramin Apandi, Asma'i Fayyadh Abd Samad, Nurfatihah Rahim, Ali Abdul Qader Mustafa, Saba Arif, Zamri Noranai

Faculty of Mechanical and Manufacturing Engineering,
Universiti Tun Hussein Onn Malaysia, 86400 Parit Raja, Batu Pahat, Johor, Malaysia

E-mail: aziemah.nz@gmail.com[1]

Abstract. Lecturer office building is particular of important buildings in Universiti Tun Hussein Onn Malaysia. Lecturers will spend their time mostly in office if no learning and teaching process in classroom. This study aims to discover potential energy saving by air conditioner temperature control setting of lecturers building at Block C15, C16 and C17 respectively. Several steps have been done to achieve the objective. A study on type of ACMV, building identification, location, the usage trend, air-conditioner quantity, and the published electrical tariff by TNB were done to familiarize with the building of study. Result was found that about 62% of electrical energy was used on Air Conditioning and Mechanical Ventilation System (ACMV). This is the most energy consumption used in lecturers building. From calculation, the monthly electrical energy consumption of lecturer buildings was near to 56.963MWh. With the proposed potential saving measures, which is air condition temperature setting, lecturers building is estimated to reduce the usage of electrical energy for about 10 to 30% per month. More potential saving could be found in future study if there is designated system that studies the energy consumption trend of the building.

1. Introduction

We are living in an increasingly shrinking world. Instant communication and the internet have seemingly dissolved time, space and cultural boundaries. The international movement of peoples has reached levels previously unheard of. Globalization has become catchphrase for our times. The atmospheric sciences have partly led and partly responded to this process. The extent of continental, hemispheric – and even global – transport of air pollution has become an issue of increasing scientific and policy concern; and nothing emphasises the fragile unity of the planet more graphically than the increasing evidence of climate change – and the portentous implications that emerge as we contemplate the possible consequences of the interaction of air pollution and climate change. Currently, buildings consume about one-third of the world's energy and the worldwide energy consumption for buildings is expected to grow from 45% from 2002 to 2025. [1-4]. The energy consumed in Malaysia is 90% in the form of electricity. If this trends continue, buildings will consume almost as much as industry and transport combined. By year 2030, 56% of the world's energy use will be from Asia. Electricity demand will grow by at least 40% by 2032. New power generation equals to nearly 300 power plants (1,000 MW) will be needed to meet electricity demand by 2030. Currently, 50% of electrical generation relies on coal and fossil fuel while 85% of greenhouse gas emissions result from energy--consuming activities supported by fossil fuels. The purpose of carrying this study is to analyze the energy consumption of Block C15, C16 and C17 per month and discover its saving potential. Solutions and recommendation are discussed in the paper, supported with measurement of approximate saving cost per month in the mosque. Energy audit is a process of reducing amount of energy input without negatively affecting the output.[8,9].

1.1. Energy Resources and Consumption
Energy is the main source to all activities. Today, demand for electrical energy had an increasing rate each day. As energy cannot be created nor destroyed, electrical energy requires energy sources to be generated, In order to understand the energy management; a study on energy resources is a must for estimating the energy consumption and proposes ways of energy saving solution.[15] The generation of electricity is using both renewable and non-renewable energy. However, the trend of world energy consumption indicates that 82% is from non-renewable energy, while the remaining 18% is from renewable energy.[16,17] Currently, there are 10 main sources of energy being used to generate electricity worldwide. The sources include solar energy, wind energy, hydrogen energy, tidal energy, wave energy, hydroelectric energy, biomass energy, nuclear power, and fossil fuels. In Malaysia, almost 80% of electricity was generated by natural gas and coal. Fossil fuel such as natural gas and goal are categorized as non-renewable energy, which cannot be replenished. Continuous usage of fossil fuel without control will finish up the resources in no time.

1.2. Building
University Tun Hussein Onn Malaysia (UTHM) is located in, Parit Raja, which is one of a town of Batu Pahat district. With the coordinates of 1.8553° N, 103.0811° E, it is approximately 4.8 kilometers or six minutes drives from Parit Raja town. There are many buildings in UTHM area such as administers building, faculties block, academics building, Masjid Sultan Ibrahim, Sultan Ibrahim Hall, residential college and lecturers building. Block C15, C16 and C17 is located near to Tun Syed Nasir Residential College and academics building. The blocks are two stories building. The total floor area of the building is 2172 m² and total number 126 rooms to cater up to 210 lecturers. Office hour for academician staff starts at 8.00 am and it ends at 5.00 pm, total sum of 8 working hours, though academic staff is not always working in their room the whole time.

1.3. Air Conditioner Temperature Setting
Air condition temperature setting point is one of the factors that contribute to total cooling load and electric energy. The lower temperature setting cause higher cooling load and higher temperature setting cause lower cooling load. Therefore, it was suggested to set the air condition temperature at higher level within the recommended range. According to the MS 1525:2014 standards, there is a standard for indoor design condition for office building as listed below.

The indoor design conditions of an air-conditioned space for comfort cooling should be as follows;

- Recommended dry bulb temperature 24 – 26 °C
- Minimum dry bulb temperature 23 ° C
- Recommended design relative humidity 50 % – 70 %
- Recommended air movement 0.15m/s – 0.50m/s
- Maximum air movement 0.70m/s

IOP Conf. Series: Materials Science and Engineering **824** (2020) 012002 doi:10.1088/1757-899X/824/1/012002

2. Methodology

The flow chart of this study is divided into 3 phases, namely Phase 1, Phase 2, and Phase 3 respectively. During phase 1, literature review has been done on several sources in order to grasp a clear conception of this study. A study on building identification, location, the usage trend, air-conditioner quantity, and the published electrical tariff by TNB were done to familiarize with the building of study. Next, Phase 2 is where data being collected and the calculation of estimated energy consumption of Block C15, C16 and C17 were done. Meanwhile in Phase 3, the analysis on potential energy saving method has been discussed, followed by the process of thesis writing. During data collection of building, the building has been studied thoroughly, so that the researcher grasps a clear understanding on the type of the building, its characteristics, and the user trend. During data collection of electrical equipment facilities, quantity of lighting and ACMV device were recorded according to their types, together with their rated power consumptions. Next, the calculation of the estimated energy consumption of lecturers building was done and the energy cost was obtained. Solution and recommendation based on the potential saving is listed afterwards, supported with simple measurement of potential saving. Some general solution includes replacement of device and increase users' awareness towards electricity usage.

3. Results and discussion

After the data collection phase, the data is presented in the table form for ease of reading and calculation. Table 1 shows the breakdown of estimated energy consumption for office building, while Figure 1 shows the table of energy consumption according to temperature setting.

Table 1: Breakdown of estimated energy consumption for office building.

Equipment	Energy Usage (kwh)	Monthly Cost (RM)	Percent (%)
Air Conditioning	42,444.00	12,223.87	74.5%
Lighting	10,487.81	3,020.49	18.4%
Office Equipment	4,032.00	1,161.22	7.1%
Total	56,963.81	16,405.58	100.00%

Table 1 shows the breakdown of the estimated electric consumption according to air condition, lighting and other equipment. Electric consumption for air conditioning was the highest (74.5%), followed by electric for lighting (18.4%) and other office equipment is the lower percent (7.1%).

Room	COOLING LOAD (BTU/H)									
Temp. set	8.00 AM	9.00 AM	10.00 AM	11.00 AM	12.00 PM	1.00 PM	2.00 PM	3.00 PM	4.00 PM	5.00 PM
70 ° F	466480.49	477666.41	480442.11	477188.44	470152.57	461611.04	464677.27	474619.03	485663.49	490470.38
71 ° F	462817.61	474003.53	476779.23	473525.56	466489.69	457948.16	461014.39	470956.15	482000.61	486807.50
72 ° F	459154.73	470340.65	473116.35	469862.68	462826.81	454285.28	457351.51	467293.27	478337.73	483144.62
73 ° F	455491.85	466677.77	469453.47	466199.80	459163.93	450622.40	453688.63	463630.39	474674.85	479481.74
74 ° F	451828.97	463014.89	465790.59	462536.92	455501.05	446959.52	450025.75	459967.51	471011.97	475818.86
75 ° F	448166.09	459352.01	462127.71	458874.04	451838.17	443296.64	446362.87	456304.63	467349.09	472155.98
76 ° F	444503.21	455689.13	458464.83	455211.16	448175.29	439633.76	442699.99	452641.75	463686.21	468493.10
77 ° F	440840.33	452026.25	454801.95	451548.28	444512.41	435970.88	439037.11	448978.87	460023.33	464830.22

Figure 1: Table of energy consumption according to temperature setting.

The figure 2 showed total cooling load was reduced about 6,153,638.4 Btu/hr per month when design temperature setting was changed from 70 °F to 77 °F. This reduced the cooling load by 5.7%, which translate into potential saving in electric energy consumption around RM697 per month with near zero cost.

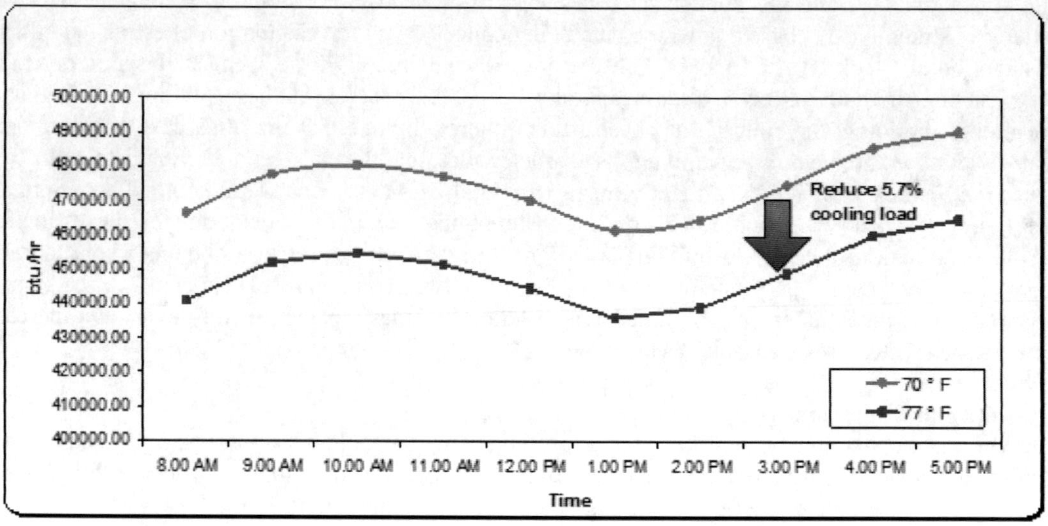

Figure 2: Graph of energy consumption according to temperature setting.

3.1 Suggestion and Recommendation

There are a few ways of solution and recommendations that were proposed in this topic. Some components or equipment replacement were included in the proposal in order to achieve cost saving objectives. Recommendation that would bring some cost saving benefits to the building has been made at the end of this study. The potential saving measures can be categorized into two categories, which are technical saving measures and manual saving measures respectively. For technical saving measures, switch off air conditioner when not use and increase its temperature in the morning and night. Meanwhile, reminder sign placement and human control are the proposed manual saving measures. Table 4 above shows the comparison between current usage and after increasing air condition temperature.

4. Conclusion

This study set out to identify building's characteristic and analyze the estimation of monthly energy consumption of Lecturers buildings (Block C15, C16 and C17) of University Tun Hussein Onn Malaysia. This study also discovers the saving potential that can be applied in lecturers building based on the data collected. However, there is no record or system that can track the electrical energy consumption of these three buildings as the electricity bill comes in bulk with several other buildings in UTHM. Hence, it is important to have estimation on the monthly electrical energy consumption to visualize the usage trend. Block C15, C16 and C17 electrical energy consumption was estimated at nearly 56.963 MWh, costing around RM 16,405.58. From these figures, the ACMV system was identified as the biggest contributor of electrical energy consumption with 74.5%, compared to others. However, the electric energy consumption usage of ACMV can be reduced by air condition temperature setting. The temperature setting that suggested to use is 77°F equivalent to 25°C. Therefore, this method not only can save electric energy consumption but in the same time in will reduce the operation cost and suitable for all building that used ACMV.

Acknowledgements
The author gratefully acknowledges the support of this work by Centre for Energy and Industrial Environment Studies (CEIES), Faculty of Mechanical and Manufacturing Engineering, (FKMP), Office for Research, Innovation, Commercialization and Consultancy Management (ORICC) and University Tun Hussein Onn Malaysia (UTHM) under Geran Penyelidikan Pascasiswazah vote number H038, and Ministry of Education Malaysia under Fundamental Research Grant Scheme vote number K091.

References
[1] Z. Noranai and M. Z. Md Yusof, "Study of energy efficiency opportunities in UTHM," *World Acad. Sci. Eng. Technol.*, vol. 77, pp. 745–751, 2011.
[2] A. Perez-Garcia, A. P. Guardiola, F. Gómez-Martínez, and A. Guardiola-Víllora, "Energy-saving potential of large housing stocks of listed buildings, case study: l'Eixample of Valencia," *Sustain. Cities Soc.*, vol. 42, no. June, pp. 59–81, 2018.
[3] Noranai, Zamri., M. N. Kammalluden, "Study of Building Energy Index in Universiti Tun Hussei Onn Malaysia," *Acad. J. Sci.*, vol. 1, no. 2, pp. 429–433, 2012.
[4] M. Brøgger, P. Bacher, H. Madsen, and K. B. Wittchen, "Estimating the influence of rebound effects on the energy-saving potential in building stocks," *Energy Build.*, vol. 181, pp. 62–74, Dec. 2018.
[5] C. Fan, Y. Sun, K. Shan, F. Xiao, and J. Wang, "Discovering gradual patterns in building operations for improving building energy efficiency," *Appl. Energy*, vol. 224, no. March, pp. 116–123, 2018.
[6] C. Alalouch, S. Al-Saadi, H. AlWaer, and K. Al-Khaled, "Energy saving potential for residential buildings in hot climates: The case of Oman," *Sustain. Cities Soc.*, vol. 46, no. January, p. 101442, 2019.
[7] M. Z. Tahir, M. N. M. Nawi, and A. Ibrahim, "Value Management (VM): A Strategic Approach for Improving Energy Efficiency," *Int. J. Supply Chain Manag.*, vol. 5, no. 4, 2016.
[8] Z. Noranai, M. H. H. Mohamad, H. Salleh, and M. Z. M. Yusof, "Energy saving measures for university public library: A case study of UTHM library," *Appl. Mech. Mater.*, vol. 660, pp. 1072–1075, 2014.
[9] L. Asere and A. Blumberga, "Energy efficiency – indoor air quality dilemma in public buildings," *Energy Procedia*, vol. 147, pp. 445–451, Aug. 2018.
[10] S. C. J. Lim and B. T. Tee, "A Preliminary Review of Building Informatics for Sustainable Energy Management," in *Journal of Physics: Conference Series*, 2018, vol. 1019, no. 1.
[11] M. Nawi, N. H. Baluch, and S. Mohtar, "Building Energy Index (Bei): a Study of Government Office Building in Malaysian Public University," no. April, pp. 192–201, 2017.
[12] T. Sciences, "Akademia Baru Building Energy Performance : A Case Study at G2 Building Akademia Baru," vol. 2, no. 2, pp. 161–169, 2018.
[13] T. Alves, L. Machado, R. G. de Souza, and P. de Wilde, "Assessing the energy saving potential of an existing high-rise office building stock," *Energy Build.*, vol. 173, pp. 547–561, 2018.
[14] J. S. Park, S. J. Lee, K. H. Kim, K. W. Kwon, and J. W. Jeong, "Estimating thermal performance and energy saving potential of residential buildings using utility bills," *Energy Build.*, vol. 110, pp. 23–30, 2016.
[15] Z. Noranai and A. D. F. Azman, "Potential reduction of energy consumption in public university library," in *IOP Conference Series: Materials Science and Engineering*, 2017, vol. 243, no. 1.
[16] V. Peyramale and C. Wetzel, "Analyzing the Energy-saving Potential of Buildings for Sustainable Refurbishment," *Procedia Environ. Sci.*, vol. 38, pp. 162–168, 2017.
[17] R. Chedwal, J. Mathur, G. Das Agarwal, and S. Dhaka, "Energy saving potential through Energy Conservation Building Code and advance energy efficiency measures in hotel buildings of Jaipur City, India," *Energy Build.*, vol. 92, pp. 282–295, 2015.

Single Parameter Engine Analysis and Performance Optimization of a Medium Duty Gasoline Engine

Mohd Faisal Hushim[1]*, Zamri Noranai[2], Daniel Lau Yan How[2], Zainul Ameerul Ikhsan Zainul Abidin[2], Mohd Azahari Razali[2], Azwan Sapit[2] and Akmal Nizam Mohamed[2]

[1]Automotive & Combustion Synergy Group (ACSG), Faculty of Engineering Technology, Universiti Tun Hussein Onn Malaysia, Pagoh Branch, 84600 Pagoh, Johor, Malaysia.
[2]Faculty of Mechanical & Manufacturing Engineering, Universiti Tun Hussein Onn Malaysia, 86400 Pt. Raja, Batu Pahat, Johor, Malaysia.

E-mail: mdfaisal@uthm.edu.my

Abstract. This paper examines various attempts to impose a single parameter analysis on the effect of intake valve opening (IVO) timings and intake runner length (IRL) technologies on the engine performance output at full load condition through an engine model representing 1.6 liter in-line 4 cylinders, 4-stroke SI engine constructed using GT-Power simulation software which is then simulated and correlated to the CAMPRO 1.6L model constructed previously with an updated crank train geometry such as cylinder stroke length and the connecting-rod length (con-rod) where comparison results of Volumetric Efficiency (VE), Brake Power (BP-kW), Brake Torque (BT-Nm) are carried out to maintain the accuracy and validity of the model to be used. This model is then used to simulate the engine performance under different sets of intake runner length (IRL) and difference sets of intake valve opening (IVO) timings. The various sets of IRL and IVO are then configured to correspond with each other to predict the most suitable optimization setups at targeted engine speeds, ranging from 2000-4000 RPM in this research to eliminate the existence of performance dip for better cycle efficiency. Results shown significant improvement through the different IVO timings and different IRL setups, as compared to the stock settings

1. Introduction

Improving the torque performance in case of naturally aspirated engines volumetric efficiency is still low which is in the range of 75-80% where in conventional engines it is optimized for a specific speed at which it gives maximum torque[1]. The effects of such engine operating parameters and their control technologies have been studied excessively by researchers and the product designers. In engines, configuration of the intake system plays also an important role on the engine performance, and there are many experimental and theoretical studies on the intake system and manifold design. In this research, analysis were carried out to investigate the design geometry of the intake subsystem on the torque output of an engine using one dimensional (1D) engine simulation program GT-Power by constructing an engine model of 1.6-liters NA, spark-ignition (SI), port-fuel injection four cylinders engine. Studies showed that longer intake manifolds give peak torque at low engine rpm & shorter engine manifolds gives peak torque at higher engine rpm[2–4]. Development of the subsystem such as intake manifold length tuning can be improved through simulation work during the engine development process[5]. The paper presents the model construction, correlation with actual test data, predicted engine performance at various IRL and IVO timing as well as their optimum configurations at specific engine speeds to determine an optimum value in order for the performance dip at 2000 rpm to 4000rpm to be eliminated.

2. Single parameter engine analysis

In this research, the effects of IRL and IVO timings on engine performances are investigated so as to find out their optimum operating values at a specific engine speeds. Using GT-Power to simulate the one dimensional (1-D) engine representing the targeted engine, which is the CAMPRO 1.6L base model requires the actual input data, representing the actual geometry of the engine[6]. Constructed model is then correlated to the model reference of CAMPRO 1.6L provided through the percentage difference in the BSFC and intake manifold VE for the validity of the model well agreed with the exact model. The model is then simulated with different sets of IRL and IVO, comparing their output performances to the stock IRL and IVO setups where 3 DOEs are carried out to determine the optimum values at specific engine speeds.

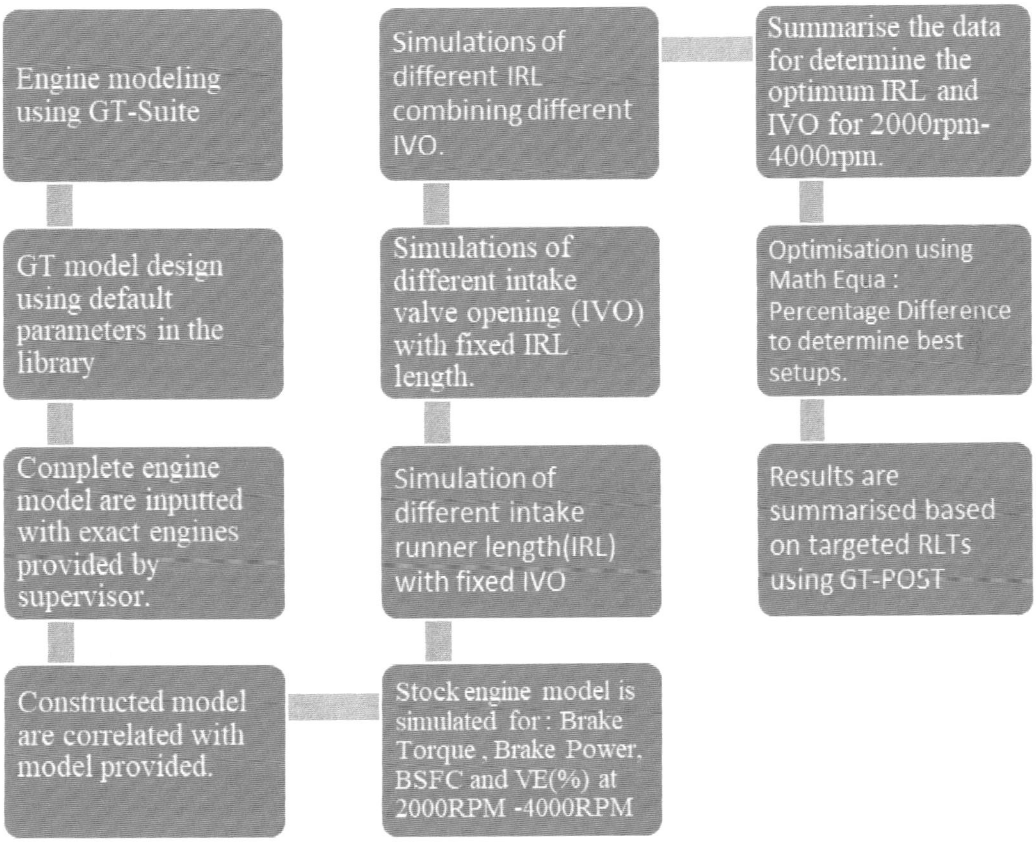

3. GT-POWER Engine Modelling

3.1 Constructed Model.

In order for the performance output data from the simulation to be as accurate as possible, it is essential to input the engine data closest to the exact value so that the results obtained and proposed optimum values from the simulation can be made valid. The model is constructed from the initial intake system until the end exhaust systems to include as much as parameters to represent the actual engine operating conditions and was represented by name CAMPROBASE1.6. The intake manifold air-box was modelled in Solidwork CAD with the exact dimensions and was discretized using GEM3D in GT-Power to obtain the exact dimensions of each sub-part in the intake system such as Inlet and outlet diameter of each pipe as well as the length are precisely defined. Non-predictive combustion model known as SI Wiebe function is chosen which imposes a burn rate as a function of crank angle (CA), represented by crank angle where 50% fuel burned and the duration of fuel burned and the burn rate is calculated from measured in-cylinder combustion pressure at various engine speeds under full load conditions. Then, the model is simulated at engine speed of 1500 to 6500 rpm under wide open throttle (WOT). After the simulation, the results of engine performance are plotted in GT-POST.

3.2 Model Correlation: STOCK vs. CAMPROBASE16.

The constructed model was correlated to the stock setups to check on the validity of the model constructed through the percentage difference (%Diff.) method, comparing the BSFC and intake manifold VE. As seen from Figure 1 and Figure 2, the brake specific fuel consumption and intake manifold volumetric efficiency of the built model. Performance difference of less than 1% confirms that it can represent the real engine, there is able to use it for other simulation purpose or predictive

BSFC	
RPM	%Diff
7000	0.018733
6500	0.023715
6000	0.021958
5500	0.008987
5000	0.008191
4500	0.015135
4000	0.003983
3500	0.009127
3000	0.006158
2500	0.009306
2000	0.024454
1500	-0.00436

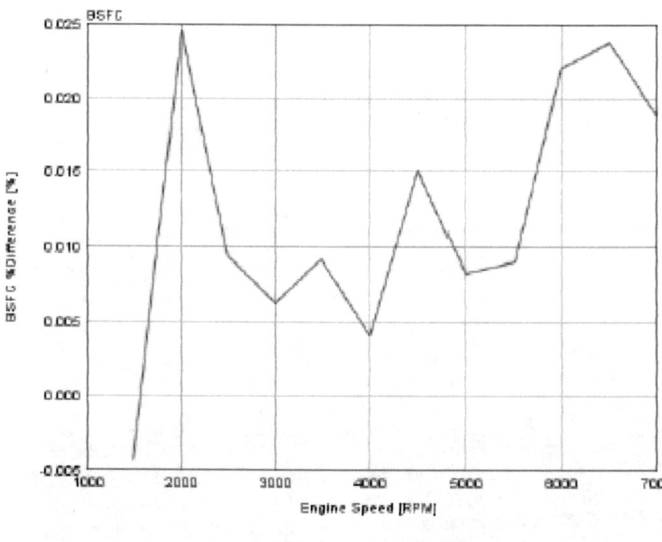

Figure 1. BSFC %Diff. Stock Model vs. Built Model.

IOP Conf. Series: Materials Science and Engineering **824** (2020) 012003 doi:10.1088/1757-899X/824/1/012003

VE intake manifold	
RPM	%Diff
7000	-0.09483
6500	-0.10286
6000	-0.09732
5500	-0.07198
5000	-0.04562
4500	-0.0386
4000	-0.02278
3500	0.030748
3000	0.073685
2500	0.043578
2000	-0.0255
1500	-0.0216

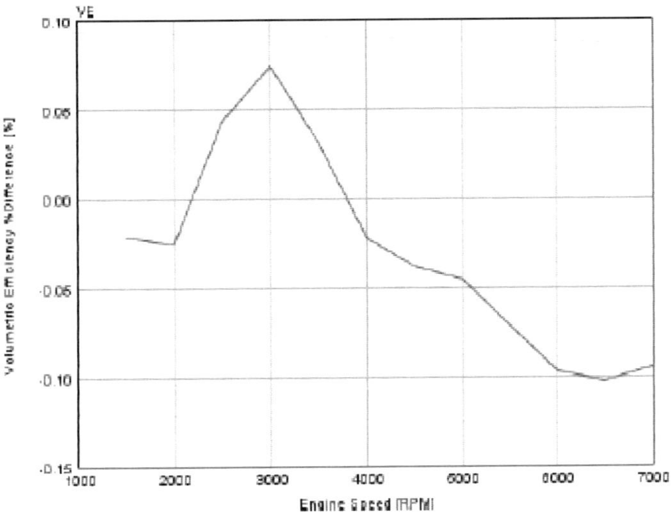

Figure 2. Intake Manifold VE %Diff. Stock Model vs. Built Model.

4. Results

4.1 1st Doe: Intake Runner Length (IRL) 350-800 mm

The stock intake runner consists of several discretized part, divided into runner 1, runner 2 and runner 3 with each length of 425.603 mm, 421.461 mm, 422.357 mm, 429.379 mm respectively. To simulate the effect of modified IRL to the performance output of the engine, these values are then modified ranging from a shorter IRL of 350 mm to a longer IRL of 800 mm with a length increment of 100 mm each after 400 mm. Figure 3 shows the overall performance output percentage difference when different sets of IRL are applied and Table 1 shows the percentage difference value comparing to the stock IRL setups.

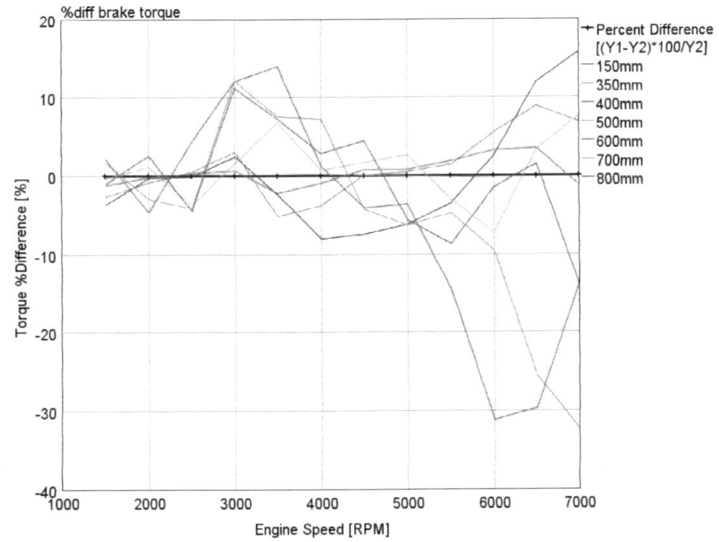

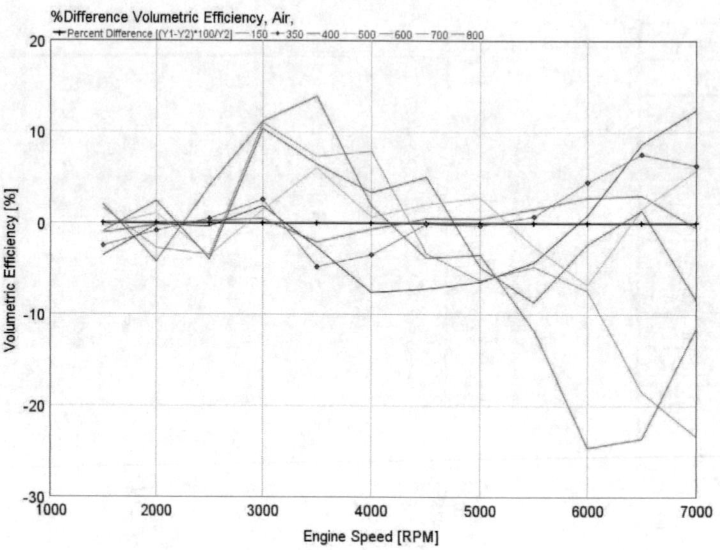

From Table 1, it is clearly seen that by increasing the IRL from the stock length, starting from the longer IRL of 500 mm to 800 mm ,noticeable improvement can be noticed at 2000 rpm to 4000rpm with an average of 1-5% of improvement obtained. However, performance shows reversed behavior after 4000 rpm where performances are significantly reduced from 5000 rpm to 7000 rpm. The shorter IRL of 150 mm, 350mm and 400mm produced significant improvement at higher engine speeds with an average improvement of 4% at higher engine speeds while suffers an average 0.5-1.7% of performance downfall at lower engine speeds. This matched with the theory where longer IRL can be used to pursue better torque performance at lower engine speeds while shorter IRL can be used to produce better performance output at higher engine speeds. Longer runner produced slightly lower BSFC at low-engine speeds with an average decrement of 1% but shoots up to an average increment of 1-5% of BSFC at the higher engine speeds ranging from 5000 rpm to 7000rpm as shown in Table 2.

Table 1. %Diff. of BT for different IRL.

%Difference BT _STOCK IVO-108°ATDC_vs_IRLs(150-800MM) based on stock model.							
(RPM)	150MM	350MM	400MM	500MM	600MM	700MM	800MM
7000	15.74581	6.861847	-1.22544	7.664782	-13.9574	-32.2732	-13.7541
6500	11.9367	8.853927	3.486987	3.311086	1.53963	-25.5108	-29.6026
6000	2.544208	5.526415	3.247986	-7.24179	-1.57997	-9.59742	-31.1654
5500	-3.54788	1.386779	1.901775	-3.0454	-8.72447	-4.76311	-14.5511
5000	-6.25443	0.477758	0.782437	2.686367	-5.65936	-6.30897	-3.6544
4500	-7.51495	0.052465	0.7143	1.727355	4.46498	-4.2747	-4.0819
4000	-8.04238	-3.80719	-1.00164	0.670356	2.892877	7.20859	1.238241
3500	-2.37102	-5.20256	-2.25969	6.762236	7.224787	7.567102	13.95668
3000	2.467428	3.077728	0.646848	1.566288	11.13094	12.1637	12.04698
2500	-0.20679	0.486565	0.293237	-3.86679	-4.54942	-4.25417	4.336029
2000	-0.32397	-0.8628	-0.21568	1.11636	2.597884	-3.00191	-4.66352
1500	-3.70537	-2.70681	-1.22027	-0.36891	-1.02284	1.727732	2.194101
Ave %Diff (2k - 4k rpm)	-1.69534	-1.26165	-0.50739	1.249689	3.859414	3.936662	5.38288
Ave %Diff (1.5k -7k rpm)	0.060612	1.178676	0.429237	0.915161	-0.4702	-5.10976	-5.64175
Ave %Diff (5k -7k rpm)	4.084879	4.621345	1.638748	0.675009	-5.67632	-15.6907	-18.5455

Table 2. Diff. of BSFC for different IRL.

(RPM)	150mm	350mm	400mm	500mm	600mm	700mm	800mm
7000	-2.91884	-0.55284	0.742928	1.787139	6.458782	13.19108	2.614866
6500	-2.93037	-1.21983	-0.4566	1.783037	-0.18804	9.360139	8.511794
6000	-1.65258	-0.95014	-0.42398	-0.4827	-0.77302	2.38278	9.511482
5500	-0.76879	-0.69368	-0.35119	-0.4522	0.100767	-0.05766	3.368701
5000	-0.17661	-0.72894	-0.31984	-0.0823	0.904797	0.001055	0.153552
4500	0.348199	-0.13795	-0.27561	-0.3264	0.591574	1.04228	0.292117
4000	0.448152	0.378964	0.302379	0.04905	0.470994	0.599005	0.599316
3500	-0.30145	0.454579	0.15632	0.693299	-0.70857	-0.22443	-0.08297
3000	-0.58081	-0.4928	-0.17147	0.147975	-0.74979	-1.01155	-0.82458
2500	-0.1135	-0.04681	-0.07303	-0.35499	0.56487	0.767953	-0.20145
2000	0.108031	0.038113	0.015469	0.097726	-0.12704	0.31354	0.484238
1500	0.188215	0.200748	0.087189	0.008488	0.064313	-0.14361	-0.19935
Ave %Diff (2K-4K)	-0.08792	0.066409	0.045934	0.126612	-0.10991	0.088904	-0.00509
Ave %Diff (5K-7K)	-1.68944	-0.82909	-0.16174	0.510596	1.300657	4.97548	4.832079

4.2 2nd Doe: Intake Valve Opening (IVO) 96°-120° ATDC

Theoretically, the IVO make impact on the engine performances output as it is closely related from the mechanism of intake manifold where the IRL determine the flow characteristics before reaching the intake port. Therefore in this research 2nd DOE, IVO is made as a single parameter that will produce different significant results if being altered, in which the timings are varied to investigate how it affects the BT and BSFC as shown in Figure 4. The IRL was made fixed at the stock length throughout this DOE. Study of the effect started from the analysis of stock IVO that is 108° ATDC. The engine performance simulated from different sets of IVO timings ranging from a retarded 96° ATDC to an advance timing of 120° ATDC where the timings are defined at the crank angle (CA) point where the intake valve opens at maximum lift, known as Maximum Opening Point (MOP) in GT-Suite.

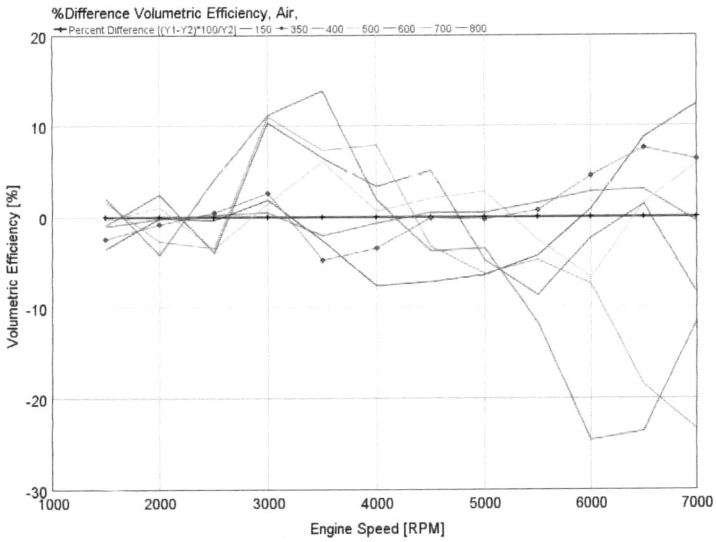

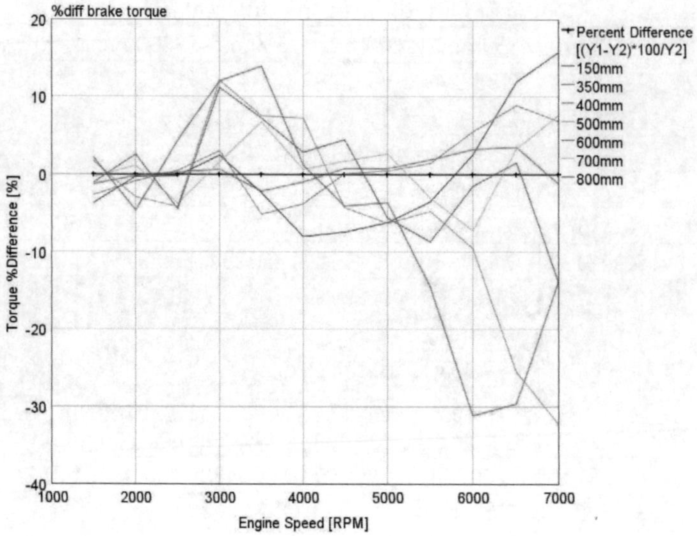

From results shown in Figure 4, it is obvious that by advancing the IVO timings from the stock 108°ATDC, the BT and VE shows significant improvements over the stock timing at higher engine speeds above 5000 rpm. However, the behavior changes when it is at lower engine speeds where decrement in BT and VE is obvious at low to mid-engine speeds from 2000 rpm to 4000 rpm. These can be seen from Figure 5 and Table 3 that the behavior shows reversed behaviors when the timing is retarded from the stock timings. In this DOE, the stock timings of 116° ATDC is said to be the most suitable timing for whole engine speeds.

However, IVO 104° can be used as an alternative in pursuing average improvement in engine speeds ranging from 2000 rpm to 4000 rpm while 120° can be chose to seek improvement in 5000 rpm to 7000 rpm as it shows average improvement of 4% in the BT and 3% of improvement in VE as shown in Table 4. IVO 116° opt to be used if average improvement in BT is seek over the entire engine speeds as shown in Table 3.

Table 3. %Difference BT :IRL_vs_IVOs(96°-120°ATDC)

%Difference BT _STOCK IRL_vs_IVOs(96°-120°ATDC) based on stock model.							
(RPM)	96 ATDC	100 ATDC	104 ATDC	108 ATDC	112 ATDC	116 ATDC	120ATDC
7000	-10.2156	-6.5423	-3.0701	0.0000	4.7854	2.6225	6.4588
6500	-10.3839	-6.6902	-3.2134	0.0000	5.2032	2.8155	7.0793
6000	-8.8446	-5.6238	-2.6382	0.0000	3.9516	2.2050	5.1921
5500	-6.6618	-4.0282	-1.7717	0.0000	1.7310	1.1675	1.7391
5000	-3.9507	-1.9784	-0.6286	0.0000	-0.4320	0.0139	-1.2634
4500	-2.4289	-0.9709	-0.1530	0.0000	-1.2828	-0.4216	-2.4833
4000	-1.4107	-0.4688	0.0050	0.0000	-1.2869	-0.4531	-2.4697
3500	-0.8275	-0.1583	0.1518	0.0000	-1.8357	-0.6654	-3.4758
3000	-1.6766	-0.5826	0.0176	0.0000	-1.8725	-0.6357	-3.6368
2500	-1.0768	-0.2212	0.1420	0.0000	-1.7323	-0.6402	-3.2225
2000	-1.0845	-0.1992	0.1850	0.0000	-2.0959	-0.7713	-3.9026
1500	-0.6255	0.1809	0.4220	0.0000	-2.7545	-1.0646	-4.9434
Ave %Diff 2k-4krpm	-1.2152	-0.3260	0.1003	0.0000	-1.7647	-0.6331	-3.3415
Ave %Diff (1.5k-7k rpm)	-4.0989	-2.2736	-0.8793	0.0000	0.1982	0.3477	-0.4107
Ave %Diff (5k 7k rpm)	-8.0113	-4.9726	-2.2644	0.0000	3.0478	1.7649	3.8412

Table 4. %Difference VE :IRL_vs_IVOs(96°-120°ATDC)

%Difference Vol.Eff _STOCK IRL_vs_IVOs(96°-120°ATDC) based on stock model.							
(RPM)	96 ATDC	100 ATDC	104 ATDC	108 ATDC	112 ATDC	116 ATDC	120ATDC
7000	-7.8373	-5.0405	-2.3799	0.0000	2.0570	3.7862	5.1666
6500	-8.0664	-5.2129	-2.5117	0.0000	2.2204	4.1257	5.6539
6000	-7.0758	-4.5075	-2.1208	0.0000	1.7851	3.2154	4.2576
5500	-5.5442	-3.3584	-1.4817	0.0000	0.9920	1.4968	1.5584
5000	-3.4191	-1.7293	-0.5631	0.0000	0.0501	-0.2799	-0.9204
4500	-2.1117	-0.8515	-0.1427	0.0000	-0.3434	-1.0494	-2.0260
4000	-1.1574	-0.3710	0.0161	0.0000	-0.3910	-1.0947	-2.0778
3500	-0.6960	-0.1246	0.1367	0.0000	-0.5745	-1.5719	-2.9565
3000	-1.5427	-0.5567	-0.0081	0.0000	-0.5328	-1.5834	-3.0804
2500	-1.0043	-0.2234	0.1159	0.0000	-0.5538	-1.4998	-2.7805
2000	-1.0328	-0.2091	0.1556	0.0000	-0.6833	-1.8573	-3.4505
1500	-0.6057	0.1492	0.3777	0.0000	-0.9616	-2.4847	-4.4474
Ave %Diff 2k-4krpm	-1.0866	-0.2970	0.0832	0.0000	-0.5471	-1.5214	-2.8691
Ave %Diff (1.5k-7k rpm)	-3.3411	-1.8363	-0.7005	0.0000	0.2553	0.1002	-0.4252
Ave %Diff (5k-7k rpm)	-6.3886	-3.9697	-1.8114	0.0000	1.4209	2.4688	3.1432

4.3 3rd Doe: IRL (150-800 mm) vs. IVO (96°-120° ATDC) for targeted engine speeds (2000-4000 rpm)

In this 3rd DOE, VE (%) output is the main focus where different sets of IRL are set to correspond with each IVO timings and the percentage difference (%Diff) method is applied to each VE results obtained to determine the most suitable setups with either the least drop in performance or show improvement in the VE output from each sets of IRL-IVO combinations. The representative from each combinations will then be analyzed for the suitability based on the highest positive difference based on the stock setups. The combinations started with the shortest IRL 150mm responding to IVO ranging from 96°-120° ATDC. The steps are then repeated up to the longest IRL 800mm until all the VE output data are obtained and then analyzed the percentage of difference over the stock setups VE values. The optimum IRL and IVO timings combinations for targeted engine speeds of 2000 rpm – 4000 rpm are determined from the overall best setups chosen from each combinations as shown in Figure 5.

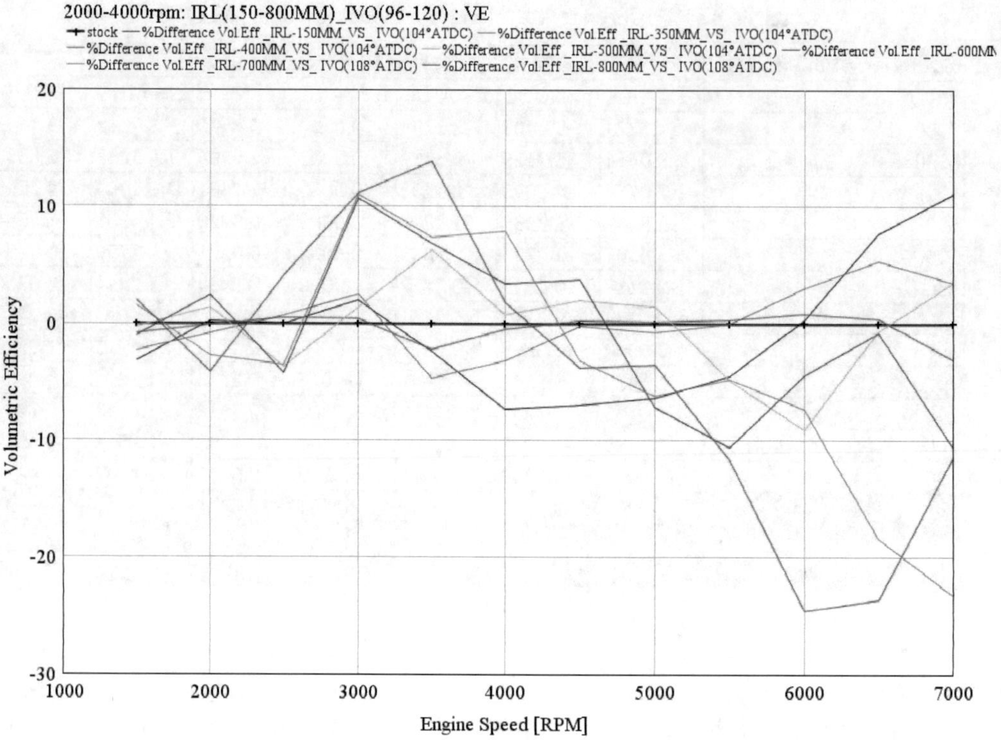

As seen from Table 5, the highest positive average %Diff by IRL 800mm explains the theory where longer IRL happens to the best setup as compared to the rest where it is obvious that in order to improve the volumetric efficiency at mid-engine speed it is essential to tune the IRL to a longer length so that maximum in-cylinder pressure can be achieved through better positive pressure difference between the intake manifold and intake port so that the oscillating wave can reach accordingly when the intake valve open while IVO 108° remains as the most suitable IVO timings in terms of CA as it produces highest positive %Diff of VE and BT at mid-engine speeds. However, IVO 104° can be considered although it lacks punch in producing better VE or BT as compared to IVO 108°.

Table 5. Best setups chosen based on the average VE %Diff. For engine speeds 2000 rpm – 4000rpm

Best Setups based on best %Difference from each IRL_vs_IVO setups based on stock model settings.

(RPM)	150MM-104°	350MM-104°	400MM-104°	500MM-104°	600MM-104°	700MM-108°	800MM-108°
4000	-7.3609986	-3.1664288	-0.5034527	0.78968686	3.439479	7.9168844	1.8447593
3500	-2.394258	-4.669395	-2.2108743	6.3525414	6.866921	7.337818	13.862206
3000	1.9594609	2.5047073	0.39381117	1.3740511	10.756241	11.052715	11.123277
2500	-0.03870736	0.6984103	0.49559128	-3.721932	-4.255058	-3.5766013	4.1260095
2000	0.26337367	-0.8462217	-0.14526978	1.346408	2.421212	-2.717544	-4.2018256
Ave %Diff (2K-4K RPM)	-1.51422588	-1.09578558	-0.39403887	1.22815107	3.845759	4.00265442	5.35088524

5. Conclusion

Through the simulating using GT-Suite, it is found that the performance dip at the targeted engine speed sourced from the sudden drop of VE which causes the BT to drop as well. Therefore, research was focused on the intake manifold runner length where different sets of IRL are implemented into the model and simulated to see the effect on the VE and BT output to propose a valid optimization sets if IRL and IVO in order to improve the performance dip at the targeted engine speeds. In the later part of the research, the different sets of IRLs are used to correspond with different sets of IVO timings (CA-deg) to determine for the most suitable setups from the combinations through the percentage difference(%Diff) method where the new data resulted from each simulations are compared to the stock setup value. Combinations that shows highest positive percentage difference is chosen to be the choice of optimization setups. The results obtained through the %Diff comparisons method satisfy the theory where long intake runner length such as 700mm and 800mm can be implemented to improve BT and VE performance at low-end to mid-end engine speeds while short intake runner length shows better improvement for the high end speeds. IVO timings of 104° and 108° are chosen as the suitable timings for the CAMPRO 1.6L as these 2 IVO shows highest positive percentage difference which indicates improvement over the stock settings across the targeted engine speeds in this research.

Recommendation

There are some of the options that could bring more accuracy in this research that is to fully model the CAMPRO 1.6L with the exact value and exact environment which is time consuming but could be consider as an option to fully investigate the affecting parameters and to come up with a better solution of optimization throughout the engine components, not just the intake subsystems and also on the engine head components where it involves a deeper review into each components of the engine in order to improve the whole engine cycle efficiency.

A dual intake runner length manifold can be implemented with characteristics such as:

- Provide the optimum length for each engine speeds for maximum performance possible.
- Significant improvement in volumetric efficiency & brake power across the whole engine speeds without any compensation of performance drop in any speeds.
- Brake Torque benefit is obtained in lower as well as higher RPM band for a more balanced drive cycle.

References

[1] G, B. J. & Jadhav, N. P. Effect of Variable Length Intake Manifold on Performance of IC Engine. **5,** 47–52 (2016).

[2] Malkhede, D. N. & Khalane, H. Maximizing Volumetric Efficiency of IC Engine through Intake Manifold Tuning. *SAE Techinical Pap. 2015-01-1738* (2015). doi:10.4271/2015-01-1738.Copyright

[3] Ceviz, M. A. & Akin, M. Design of a new SI engine intake manifold with variable length plenum. *Energy Convers. Manag.* **51,** 2239–2244 (2010).

[4] Malkhede, D. N. & Khalane, H. Maximizing Volumetric Efficiency of IC Engine through Intake Manifold Tuning. in (2015). doi:10.4271/2015-01-1738

[5] Safari, M., Ghamari, M. & Nasiritosi, A. Intake Manifold Optimization by Using 3-D CFD Analysis. in (2003). doi:10.4271/2003-32-0073

[6] Said, M. F. M., Latiff, Z. A., Saat, A., Said, M. & Abidin, S. F. Z. Analysis of variable intake runner lengths and intake valve open timings on engine

Acknowledgments

The authors would like to express their gratitude to University Tun Hussein Onn Malaysia (UTHM) for providing the financial grant vote no. H106 (Tier 1) and facilities to carry out this research.

Study on Behavior of Water Treatment Pump Before and After Modification using Finite Element Modal Analysis

Mohamed Firdous Bin Fasal Mohamed[1*], Nor Azali Bin Azmir[1*]

Faculty of Mechanical and Manufacturing Engineering, Universiti Tun Hussein Onn Malaysia (UTHM), Batu Pahat 86400 Johor, Malaysia.

Abstract. The aim of this study is to conduct a finite element analysis towards an existing water treatment pump and modification in order to reduce resonance. The application of water treatment pump is a way to inhale dirty water from the lower pressure to the high pressure area. Improvement and study about water treatment pump system to reducing the resonance with several modifications on the pump structure. Therefore, the pump will be designed to meet the requirement of the system. The basic concepts of available water treatment pumps were studied thoroughly. Designing with the actual measurements, the next step was to analyze the design using Finite Element Analysis (FEA) for simulations of natural frequency, static and dynamic analysis via Solidworks2017. It is to make sure that the system can function without any failure and to determine the vibration level due to resonance phenomenon. Result for modal analysis shows that the excitation frequency which experienced by the pump casing is 47.57 Hz and the natural frequencies are above 500 Hz which is too high. After adding damping system to the pump casing, the natural frequency reduced to 6.9286Hz, 20.254 Hz, 9.477 Hz, 46.29 Hz and 71.819 Hz which in turn reduces the vibration. For rotary screw, natural frequencies were reduced when the material stiffness has been decreased from 74,000 N/m^2 to 72,400 N/m^2.

1. Introduction

There are several types of pumps used in industries such as centrifugal, propeller, turbo and positive displacement. Water treatment pump plays an important role in industries like petrochemical, to remove slurry from petroleum. Vibration and noise are the most common pump malfunctions caused by the interaction of multiple factors and various reasons[1]. Excessive level of vibration may cause severe damage to pump bearing, shaft, impeller and other relevant parts in the pump. In order to increase lifespan of the pump, several modifications can be made. A modal analysis on the water treatment pump will be carried on before the modifications takes place[2].

Based on the research, best pump modification could be obtained with these steps. The existing water treatment pump must be drawn in Solidworks2017 with its exact dimensions. Any modifications such as loosening structure, stiffening structure, adding mass, installing tuned absorber is recommended to be applied in order to prevent resonance. However, it is needed to be ensured that the modification can well reduce the vibration effectively and safely. Therefore, to conduct Finite Element Analysis (FEA), so best modifications can be properly implemented. This work is started by collecting data such as technical drawing of pump and structure, specification of motor and pump (including mass, RPM, etc.), and material specification of structure and pump[3]. It is necessary to include exact dimensions and parameters in order to get the actual pump based results. Afterwards, drawing and modeling of existing structure are carried out. Then by using Solidworks2017, several types of modification such as loosening structure, stiffening structure, adding mass, or installing tuned absorber are performed. Solidworks2017 is one of the best software to design and analyze products in 3 dimensions. Many types of pump designs can produce by using this software in shorter period.

The general objectives of this research are to carry out modal analysis by using Finite Element Analysis to identify the resonance in water treatment pump. With the help of Solidworks2017 software, any mitigation such as loosening structure, and/or stiffening structure, adding mass, installed tuned absorber is to be applied in order to prevent resonance. The desire modifications can be well reduced the vibration effectively and safely. Finally to obtain the best modification of water treatments pump with less resonance phenomenon.

The natural frequencies and mode shapes of the pump are the modal parameters inherent in the system. These modal parameters affect the dynamic response and transmission of vibration. The modal parameters can be obtained by finite element analysis and modal analysis.

2. Finite Element Analysis

Finite element analysis is a computerized procedure for the analysis of structures and other continua. Rapid engineering analyses can be performed because the structure is represented (modeled) using the known properties of standard geometric shapes, i.e., finite elements. Efficient, large, general-purpose computer codes now exist with appropriate matrix assembler routines and equation solvers for calculation of the following structural properties:

a) Static displacement and static stress.
b) Natural frequencies and mode shapes.
c) Random forced response, random dynamic stress.

General purpose finite element codes such as NASTRAN, ANSYS, SAP, ADINA, Solidworks2017, etc., are programmed to develop and solve the matrix equation of motion for the structure:

$$[M]\{\ddot{u}\}+[C]\{\dot{u}\}+[K]\{u\}=\{F\cos\omega t\} \qquad (1)$$

The model details are entered by the analyst in a standardized format. The computer then assembles the matrix equation of the structure. The first part of the solution (1) to a given problem is to solve the matrix equation;

$$[M]\{\ddot{u}\}+[K]\{u\}=0 \qquad (2)$$

for the free vibrations of the structure. The solution to equation (2) gives the natural frequencies (eigenvalues) and the undamped mode shapes (eigenvectors). These parameters are the basic dynamical properties of the structure, and these parameters are needed for use in subsequent analysis for dynamic displacements and stresses [4].

Finite element analysis in conjunction with the high-speed digital computer permits the efficient solution of large, complex structural dynamics problems. As the majority of structural dynamics problems are linear and can be solved in the frequency domain using a modal transformation as noted above, subject to certain simplifying assumptions concerning the nature of damping.

The finite element method therefore offers a very efficient procedure for the calculation of complex linear structures under a variety of dynamic excitation conditions, and under environmental conditions, which may include temperature effects and entrained fluid effects. Where the structure is nonlinear, modal testing may still be used (with caution) to estimate initial values for mass, stiffness, and damping parameters, which can then be modified to suit more advanced structural models [5].

Although most linear structural dynamics problems may now be solved accurately and economically, it is still costly to solve most non-linear problems. For such cases a solution strategy must usually be developed on a case-by-case basis. In such instances the structural geometry and elasticity may be needed in considerable detail in the input data, and the formulation time for such cases may be significant unless suitable pre-processors are available within the code. Efficient computation of such recurrent components has been undertaken by special finite difference procedures, but sub-routines to undertake such computations are not yet in widespread use [6].

3. Modal Analysis

An analysis which provides an indication of the limits of the response of a particular system is known as modal analysis. Basically every object can naturally vibrate with presence of an internal frequency (or resonant frequency) in the object. It is also the frequency where the object will allow transmission of energy from one form to the other with minimal loss – here vibrational to kinetic. It is better to know the structure of an object can behave unevenly with these frequencies [7].

Natural frequencies and mode shapes of structures can be identified by using formalized method which uses in modal testing. Modal test produces results such as various natural frequencies, mode shapes, and related data of the structure. In modal testing, data were collected using effectual curve fitting routines on the digitized input signals. The results will be animated which displays impedance plots and mode shapes. There were some cases where an earlier precise modal analysis have prevented loss to lives and property, for instance in case of an earthquake.

First, a modal test provides the most rapid and effective procedure available for the acquisition of data on the dynamic properties of a structure. Second, modal analysis is an effective analytical procedure for the solution of large sets of structural dynamics equations because it reduces coupled matrix equations (which must otherwise be solved by some iterative procedure) to a set of independent linear equations, each with the well-known closed-form solution given above. Modal solutions can therefore be obtained directly, without further numerical operations.

These solutions are then re-combined to form the complete solution to the structural response problem in question. It should here be noted that solutions to harmonic, transient, and random forced vibration problems can all be obtained using this modal analytical procedure [8].

The main assumption involved in the acquisition of this information is that the structural system is linear, i.e., structural displacements are directly proportional to applied loads. In practical structures this condition is not always met. Structural systems may be non-linear to some degree, due to those causes listed below. Nonlinearities complicate the extraction of modal data and, where the effect is strong; it may invalidate the results obtained by linear analysis.

Non-linear effects complicate the analysis and tend to introduce errors into the data reduction and curve-fitting estimates of natural frequencies. Such results cannot always be adequately represented by a linear analysis, because the properties change according to the magnitude of the applied load [9].

4. Methodology

Good design requires both analysis and synthesis. Typically, approach complex problems like a design decomposing the problem into manageable parts. It is needed to understand how the part will perform in service and must be able to calculate as much about the part's behaviour or possible by using the appropriate disciplines of science, engineering and computational tools [10].

Drawing and modeling of each modification is performed. Thus, it can be estimated which modification could shift the natural frequency far away from operating speed. After the best modification is obtained, static, dynamic and modal analysis will be conducted. Hence, it can be ensured that the modification is acceptable in material strength point of view, so it can be proposed to be implemented.

4.1 Measurements

Reference to redraw the water treatment pump and motor has taken from specification stated by the previous researcher. The measurements of water treatment pump which will be developed as follow:

IOP Conf. Series: Materials Science and Engineering **824** (2020) 012004 doi:10.1088/1757-899X/824/1/012004

Table 1. Measurements of water treatment pump.

Parameters	Dimension (mm)
Length	441
Height	183
Width	125

4.2 Preliminary Design

After considering the lowest reading on static analysis, the best design of pump will be selected among these 3 designs which is Design 1, Design 2 and Design 3 All three designs have different types of basement. Design 1 has two 'T' shape basement. While Design 2 has one 'T' shape basement attached in the middle. Lastly Design 3 has solid basement which attached from one end to another. The selected design will undergo analyses like static, dynamic and modal analysis using Solidworks2017. Based on the results in Table 2, the best design for pump casing can be chosen was Design 1. By comparing the other two designs, Design 1 has the lowest reading on static stress which is 8.317 MPa compared to Design 2 & Design 3 where 35.155 MPa and 9.911 MPa respectively. The design which has lower static stress value is tend to last longer compared to design with higher value of static stress.

Table 2. Selection of pump casings.

Static Analysis for Pump Casings				
Plot Type		Design 1	Design 2	Design 3
Static stress N/mm² (MPa)	Highest	8.317	35.155	9.911
	Lowest	0.001	0.001	0.001
Static displacement (mm)	Highest	0.012	0.180	0.01
	Lowest	0.000	0.000	0.000
Static strain	Highest	2.24×10^{-5}	1.23×10^{-4}	2.607×10^{-5}
	Lowest	6.15×10^{-9}	4.885×10^{-9}	5.608×10^{-9}

4.3 Pump Modifications

There were two modifications have been made onto the water treatment pump. First, pump casing was added with 2 springs at the bottom. Those springs act as a vibration damper in order to reduce the amount of vibration. Second modification was made on the rotating screw. The material of rotating screw was changed with material which has less stiffness. This will tend to reduce the natural frequency of rotating screw. Figure 2 shows the parameters applied on pump casing such as fixture, pressure, spring stiffness and gravity.

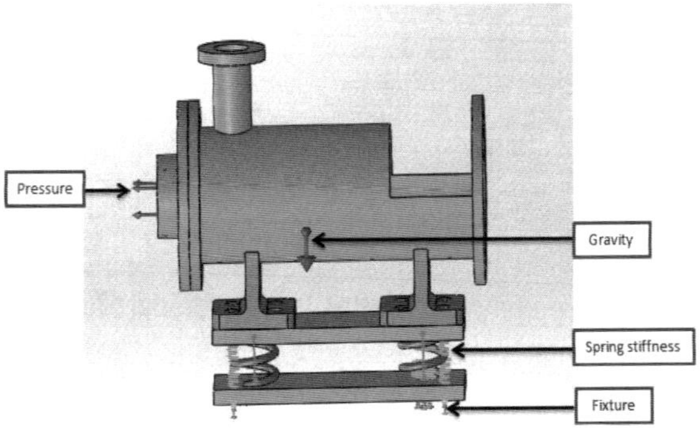

Figure 1. Parameters applied on pump casing.

IOP Conf. Series: Materials Science and Engineering **824** (2020) 012004 doi:10.1088/1757-899X/824/1/012004

There were two components of product will be analyzed which are pump casing and rotary screw. For the pump casing and rotary screw, the load of 10 kN/m^2 is put on the pressure area. Spring stiffness of 12,500 N/m was added to the pump casing [11]. Fixtures were charged at the bottom of base. Gravitational force of 9.81 m/s^2 was added vertically. Due to the whole pump structures were made of steel, the damping ratio equal to 2% was added in the analysis [12].

4.4 Finite Element Analysis
Based on Figure 2, von Mises stress indicates 6.174 MPa on pump casing which is lower than the previous unmodified pump casing which is 8.317 MPa. The springs below experience more von Mises stress compared to pump casing. The highest value of stress in this analysis is 142.742 MPa.

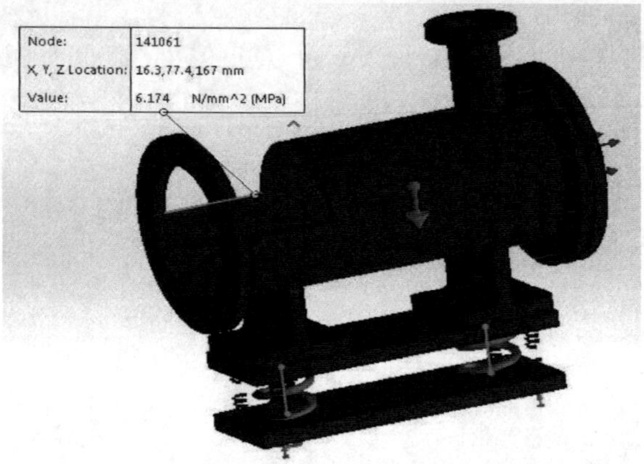

Figure 2. Static stress of modified pump casing.

According to Figure 3, von Mises stress indicates 5.156 MPa on rotary screw which is lower than the previous unmodified rotary screw which is 5.320 MPa. Moreover, this modified rotary screw has von Mises stress which is below the yielding strength 345 MPa.

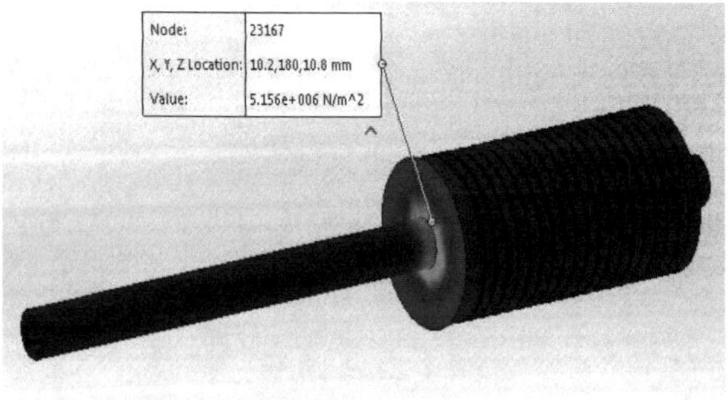

Figure 3. Static stress of modified rotary screw.

5. Result and Discussion
The analysis in Solidworks2017 is about to know the stress, displacement, strain, mode shape and natural frequency when parameters such as pressure, gravitational force and other related loads are applied.

Based on Table 3, results obtained for static analysis on pump casing can be concluded as modified pump casing has small amount of von Mises stress of 6.174 MPa experienced on its structure compared to existing pump casing which is 8.317 MPa. So that, modified pump casing tend to have less vibration compared to existing pump casing. Meanwhile for rotary screw, can be concluded as modified rotary screw has obtained lower von Mises stress of 5.156 MPa compared to existing rotary screw which is 5.320 MPa. So that, modified rotary screw tend to have less vibration compared to existing rotary screw.

Table 3. Comparison results for static and dynamic analysis.

Static Analysis					
Plot Type		Existing pump casing	Modified pump casing	Existing rotary screw	Modified rotary screw
Static stress N/mm² (MPa)	Highest	8.317	6.174	5.320	5.156
	Lowest	0.001	0.000	0.004	1.973×10^{-5}
Static displacement (mm)	Highest	0.012	0.431	0.376	0.011
	Lowest	0.000	0.000	0.000	0.000
Static strain	Highest	2.24×10^{-5}	1.847×10^{-5}	5.14×10^{-5}	8.940×10^{-5}
	Lowest	6.15×10^{-9}	9.672×10^{-10}	8.35×10^{-8}	6.281×10^{-10}
Dynamic Analysis					
Plot Type		Existing pump casing	Modified pump casing	Existing rotary screw	Modified rotary screw
Dynamic stress N/mm² (MPa)	Highest	81.122	0.631	35.64	31.31
	Lowest	0.017	0.004	0.877	0.9095
Dynamic displacement (mm)	Highest	0.022	1.408	0.041	0.041
	Lowest	0.000	0.000	0.000	0.000

Results shown on Table 3 for dynamic analysis proves that modified pump casing experience less vibration compared to existing pump casing. Von Mises stress on modified pump casing indicates only 0.631 MPa which is very much lower compared to existing pump casing, 81.122 MPa. Then by referring to the results for rotary screw, it can be concluded as modified rotary screw has small amount of von Mises stress experienced on its structure which is 31.31 MPa compared to existing rotary screw, 35.64 MPa. So that, modified rotary screw tend to have less vibration compared to existing rotary screw.

According to graph shown in Figure 4, the value of natural frequency dropped drastically after adding vibration damper (springs) at the basement of pump casing. These springs will isolate vibration in the pump casing by absorbing the energy in one way and release it in another. A good vibration damping system will lower the natural frequency of pump casing below excitation frequency. The excitation frequency which experienced by the pump casing is 47.57 Hz and the natural frequencies are above 500 Hz which is too high. After modification have made, the natural frequency reduced to 6.9286 Hz, 20.254 Hz, 9.477 Hz, 46.29 Hz and 71.819 Hz. The new natural frequencies at mode shape 1, 2, 3 and 4 are below the excitation frequency 47.57 Hz which in turn reduces the amount of vibration.

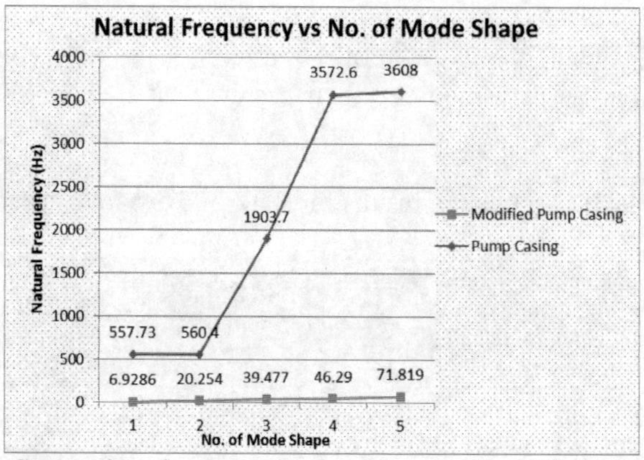

Figure 4. Comparison between pump casing and modified pump casing.

Mass= 4.56358 kg
Volume=5.8572106×10^{-4} m^3
Stiffness=1×10^{10} N/m^2 or 10,000 MPa

Mass density; m $= \dfrac{\text{Mass}}{\text{Volume}}$

$= \dfrac{4.56358\ \text{kg}}{5.8572106 \times 10-4\ \text{m}^\wedge 3}$

$= 7791.39\ \text{kg/m}^3$

Natural frequency; ω

$= \sqrt{\dfrac{k}{m}}$

$= \sqrt{\dfrac{1 \times 10^\wedge 10\ \text{N/m}^\wedge 2}{7791.39\ \text{kg/m}^\wedge 3}}$

= 1132.90 rad/sec

Conversion of radian per second to hertz:

$\omega = 2\pi v$

$\dfrac{1132.90}{2\pi} = v$

v = 180 Hz

The results obtained for natural frequency via manual calculation is too high (180 Hz) compared with results obtained through simulation. It is an acceptable value because when the number of mode shape increases, the value of natural frequency also increases.

Natural frequencies of modified rotary screw which is shown in Figure 5 are lower compared to unmodified rotary screw. Natural frequencies were reduced when the material stiffness has been decreased from 74,000 N/m^2 to 72,400 N/m^2. According to the natural frequency formula, when material stiffness was decreased or material mass was added, natural frequency will decreased.

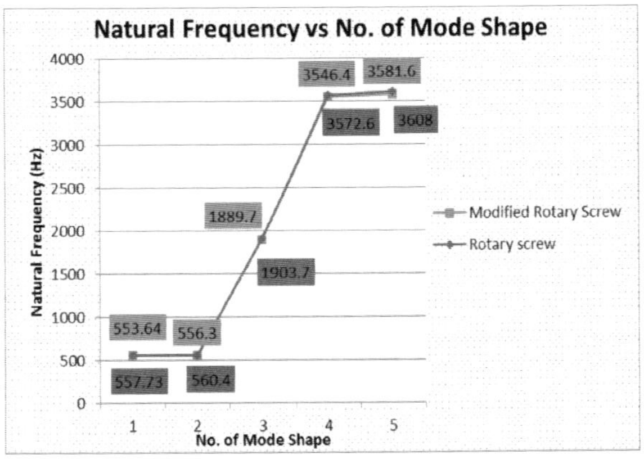

Figure 5. Comparison between rotary screw and modified rotary screw.

$$\omega\downarrow = \sqrt{\frac{k\downarrow}{m}}$$

Although the natural frequency of modified rotary screw is still higher than excitation frequency (47.57 Hz), the values of natural frequency has been successfully reduced which means the vibration level is also reduced.

6. Conclusion

This study is to carry out modal analysis by using Finite Element Analysis to identify the resonance in water treatment pump using Solidworks2017. Then to obtain the best modification of water treatment pump with less resonance phenomenon. In order to reduce resonance, modification such as adding damping system (springs) and changing material had been implemented.

The simulation analysis showed in detail about static analysis, dynamic analysis and modal analysis for both existing and modified water treatment pump. The obtained data proves that the resonance can be reduced by adding damping system and also by changing material of the body structure. It is hoped in reality, these kinds of modifications will help to reduce vibration and resonance in the water treatment pump. By modifying it, the lifespan tend to increase. Besides that, no need to frequently do preventive maintenance and it will lead to reduce the unnecessary maintenance cost.

The recommended modifications to reduce vibration on a pump are:

1. Adding vibration damper like springs at the base of pump. To obtain a better vibration reduction rather than using spring it is recommended to use rubber, polyurethane or sorbothane. Comparing to spring these 3 materials can isolate and damp vibrations.
2. Reducing the material stiffness or adding material mass.

References

[1] N. Zhang, M. Yang, B. Gao, and Z. Li (2015), "Vibration characteristics induced by cavitation in a centrifugal pump with slope volute," Shock Vib., vol. 2015.

[2] P. AmitKumar, K. K. Jain, R. KDave, and A. P. A. Choudhary, "CFD analysis of centrifugal pump impeller having different exit blade width, exit diameter and trailing edge blade angle to enhance performance," 2017.

IOP Conf. Series: Materials Science and Engineering **824** (2020) 012004 doi:10.1088/1757-899X/824/1/012004

[3] D. Zindani, A. K. Roy, and K. Kumar, "Vibration Analysis of Mixed Flow Pump Impeller Blade Designed Using Mean Stream Line Method," in *Applied Mechanics and Materials*, 2016, vol. 852, pp.476–482.
[4] S.S.Rao, The finite element method in engineering. Butterworth-heinemann,2017.
[5] S. H. Crandall and W. D. Mark, Random vibration in mechanical systems. Academic Press, 2014.
[6] N. F. Rieger, "The Relationship Between Finite Element Analysis and Modal Analysis," p. 11,1986.
[7] G. Hearn and R. B. Testa, "Modal analysis for damage detection in structures," *J. Struct. Eng.*, vol. 117, no. 10, pp. 3042–3063, 1991.
[8] J. N. Xia and H. W. Song, "Modal parameter identification of structure under base excitation using vibration test data," *Proc. Inst. Mech. Eng. Part G J. Aerosp. Eng.*, vol. 231, no. 8, pp. 1428–1450, 2017.
[9] K. Taira *et al.*, "Modal analysis of fluid flows: An overview," *AIAA J.*, pp. 1–29, 2017.
[10] W. Chapman, *Engineering modeling and design*. Routledge, 2018.
[11] Bao, H., Hatzor, Y. H., & Huang, X. (2012). A new viscous boundary condition in the two-dimensional discontinuous deformation analysis method for wave propagation problems. *Rock mechanics and rock engineering, 45*(5), 919-928.
[12] Bilbao, A., Avilės, R., Agirrebeitia, J., Ajuria, G.,(2006) Proportional damping approximation for structures with added viscoelastic dampers, Finite Elementsin Analysis and Design, 42, 497-502.

IOP Conf. Series: Materials Science and Engineering **824** (2020) 012005 doi:10.1088/1757-899X/824/1/012005

Preliminary of finite element analysis (FEA) investigation on the stent expansion

Zulfaqih Lazim, Al Emran Ismail[1], Ishkrizat Taib
Faculty of Mechanical and Manufacturing Engineering, Universiti Tun Hussein Onn Malaysia (UTHM), Batu Pahat 86400 Johor, Malaysia.

E-mail : emran@uthm.edu.my / al_emran@hotmail.com

Abstract. The requirement of stent expansion is important to stent deployment on the inner of the defected blood vessel. This defected blood vessel decreases in lumen diameter size because of accumulation of fat or plaque and caused the reduction of blood flow and oxygen supply to overall body. According to clinical guidelines, blood vessel that blockage by 70% need to be stenting. Stent expand past it elastic limit and plastically deform until the targeted diameter is achieved. Explicit dynamics Finite Element Analysis (FEA) was used in this paper to analyse the stent expansion process. The knowledge of stent radial displacement could prevent the injury of inner surface of artery while understanding on longitudinal foreshortening that aided the stent placement on targeted artery location. The controlled variable was geometry, material and boundary condition of stent. The manipulated variable was the speed of pressure application. The outcomes of this stent expansion simulation are maximum radial expansion, the uniform of radial displacement and longitudinal foreshortening. This simulation results help to improve stent installation on human body for balloon expandable stent and increase the understanding of explicit dynamics structure simulation of finite element in biomedical application.

1. Introduction

Stent is metal tube mesh structure designed to regain more blood flow at blockage blood vessel area. Mostly the blockage of 70% from original blood vessel diameter required a stent to be implanted to open up the lumen area. The stent is crimped inside catheter and guided to the location of stent placement through inside of the blood vessel, the stent then expands to the required diameter to regain normal blood flow. The increasing of fatty substances contributes to total blockage of blood vessel and lead to heart attack [1]. Therefore, it's become normal nowadays for medical doctor to treat blockage artery with stent mechanisms [2-3]. Another method of regaining normal blood flow from blockage artery is through bypass open surgery. Traditionally angioplasty is used to open up blockage artery by inflating tiny balloon in coronary artery [4]. However, re-narrowing of artery occurs after the angioplasty procedure at the defected artery location. Stent is introduced to overcome this problem and hold artery based on radial strength after nonlinear stent expansion. The stent will not recoil because it expands to plastic deformation region since it is experienced a permanent shape. However, some structural damages can be happened during the expansion process if the stent is not properly designed. In this paper, a preliminary investigation is conducted to design and analysis of the stent performances

[1] To whom any correspondence should be addressed.

under associated loading. ANSYS finite element software is used to analyse the stress behaviour occurs during the expansion process. Then, radial displacement of the stent is explored and discussed.

2. Geometrical model, material and boundary condition

Stoeckel et al. [6] estimated that more than 100 of stent design model exist in biomedical device market. There are many aspects of design variability for example in term of strut surface area, geometry, fabrication and material of the stent. The stent geometrical model of FEA simulation in this study was adopted from stent geometrical model by Chua et al. [7]. The stent model consists of 3mm outer diameter, 2.9mm inner diameter, 20mm in length and 66 slots. The dimension and slot arrangement are visualized in Figure 1. The cross-section shape of the model stent strut was chosen as rectangular shape. Figure 1(a) shows a unit cell of stent where the distance between two horizontal lines is 0.214 mm and the gap between struct is 0.40 mm. On the other hand, Figure 1(b) reveals the complete in-plane stent design before it is wrapped over the solid cylindrical bar.

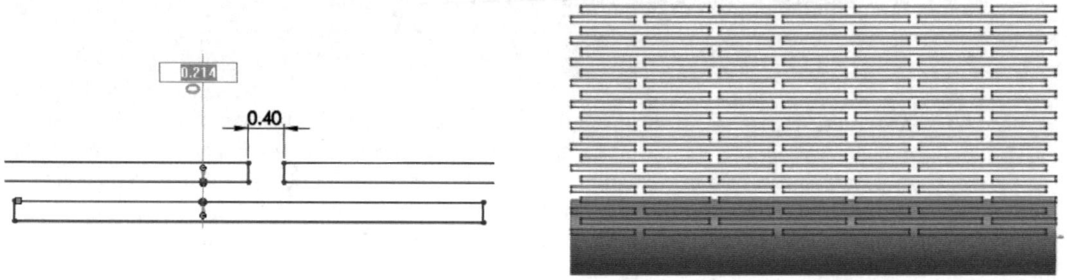

(a) (b)

Figure 1: (a) Strut's bridge and width dimension 0.400mm and 0.214mm on 2D drawing, (b) Stent wrap feature in SOLIDWORK for 3D stent model generation.

The slot or cell shape is also rectangular shape when view normal to stent surface. This stent design having similar size of cell that distributed evenly on stent surface area. Full stent surface area is 101.71mm^2 while full cylinder surface area without stent cells is 188.4956mm^2. This stent design having 53.96% surface area from full cylinder surface area with the same diameter and length. Stent material mostly made from stainless steel. It was least expensive material compare to other stent material type such as Nitinol. Stainless steel provides good radial strength for stent material [8]. The stent material assign in this Finite Element model for FEA stent expansion simulation was stainless steel 304 (SS304). Hiramoto et al. [9] also used SS304 material on stent analysis of the effect of magnetic resonance imaging. The SS304 material assignment consists of density 7.86kgmm-3, young's modulus 193MPa, Poisson's ratio 0.27, yield strength 207MPa and tangent Modulus 692MPa. A bilinear isotropic hardening on plasticity was assumed as stent material model. Symmetry by half model was choose for boundary condition for simulation of stent expansion.

Chua et al. used half stent symmetry [7] model and quarter stent symmetry model [10] for boundary condition. Observing from this two works [7 & 10], the symmetry is used for stent model while number of planes are selecting based on geometrical cell arrangement of the stent geometrical model. The selection of symmetry depending on orientation of stent bridge and strut arrangement together with number of cells. The number of cells determine the surface area of stent. Figure 2 show half stent model by symmetry in ANSYS design modeler for FEA. It is showed the stent expansion process in blockage artery. Figure 2 reveals the geometry of the stent used in this study. Figure 2(a) shows the complete or full stent model consisting of 66 slots with 0.5mm strut thickness. Since the model is symmetry, only half of the model is used as shown in Figure 2(b). This symmetry control stent simulation to expands radially and shrink by longitudinal direction on stent expansion process. However, a proper boundary condition should be aligned with stent geometrical design to avoid inaccurate FEA. Sensitivity analysis was conducted to find proper meshing size [11-12] for further stent expansion simulation. The comparison was made between nine simulations with different in mesh body sizing. Other parameters treated as constant variable such as geometry, material assignment and speed pressure application (SPA). The material used for this sensitivity analysis was SS304, while speed pressure application taken as 400MPa/ms. The simulation for sensitivity analysis objective to find suitable mesh size for further stent expansion simulation. Figure 3 shows the enlarged area of the stent when the model in Figure 2 is meshed using ANSYS finite element software. Figure 4 shows the expansion inside the blood vessel governed by ballooning mechanism.

(b) (c)

Figure 2: (a) Stent deployment in defected artery, (b) 3D stent model, and (c) Symmetry half stent boundary.

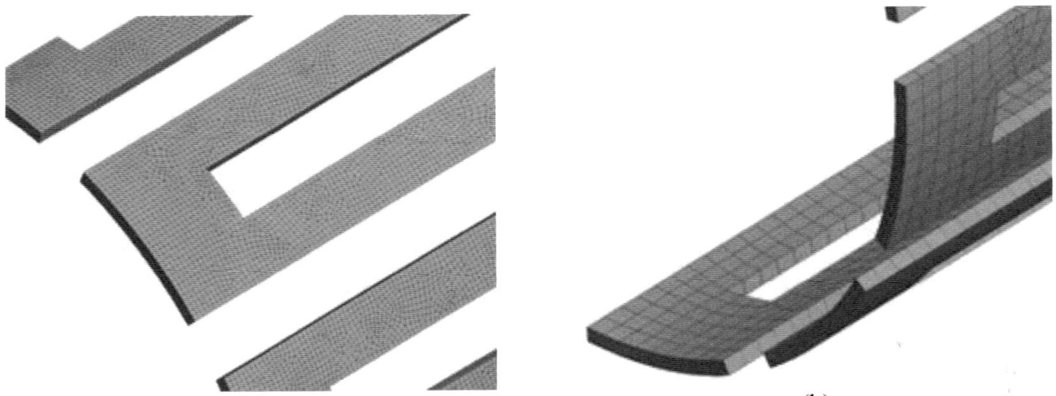

(a) (b)

Figure 3: (a) Mesh by 0.02mm body sizing consist of 411942 elements, and (b) Mesh by 0.075mm body sizing consist of 9854 elements.

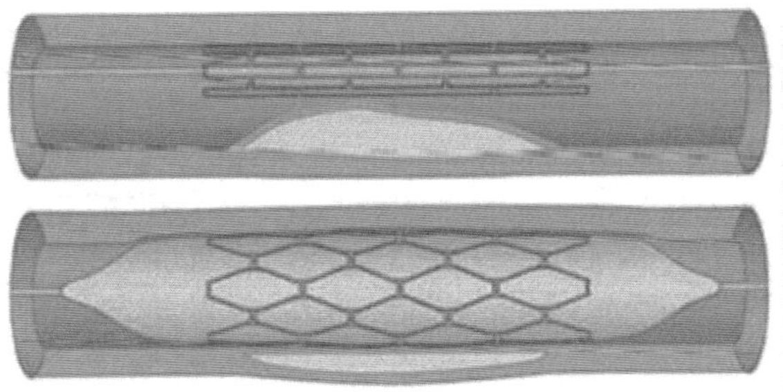

Figure 4: Stent expansion mechanisms inside the blood vessel.

Figure 5 show the visualization of mesh for sensitivity analysis, (a) for body sizing consist of 411942 elements, and (b) for 0.075mm body sizing consist of 9854 elements. The 411942 elements take simulation run time by 854.685 minutes approximately 14 hours for this sensitivity analysis study. In FEA, quality and cost are important consideration for simulation work, quality in term of accuracy required more mesh element that increase complexity however it cost more simulation time [13]. Makino et al. [14] studied simulation of car model crash analysis on the effect of fine mesh model and produced a 10-million shell elements car model to achieve the good accuracy in crash analysis. Table 1 shows element size as an input of simulation and simulation time, pressure input stent expansion and percentage error as an output of the simulation.

Table 1: List of sensitivity analysis data input and output.

No	Element Size (mm)	Number of elements (x1000)	Simulation time (ms)	Pressure Input (MPa)	Percentage Error (%)
1	0.020	411.942	0.026703	10.6812	0.00
2	0.030	123.252	0.028219	11.2876	5.37
3	0.040	68.398	0.029391	11.7564	9.15
4	0.050	21.582	0.027531	11.0124	3.01
5	0.060	15.855	0.027907	11.1628	4.31
6	0.075	9.854	0.027315	10.9260	2.24
7	0.100	6.662	0.029439	11.7756	9.29
8	0.500	0.685	0.032396	12.9584	17.57
9	0.750	0.420	0.027917	11.1668	4.35

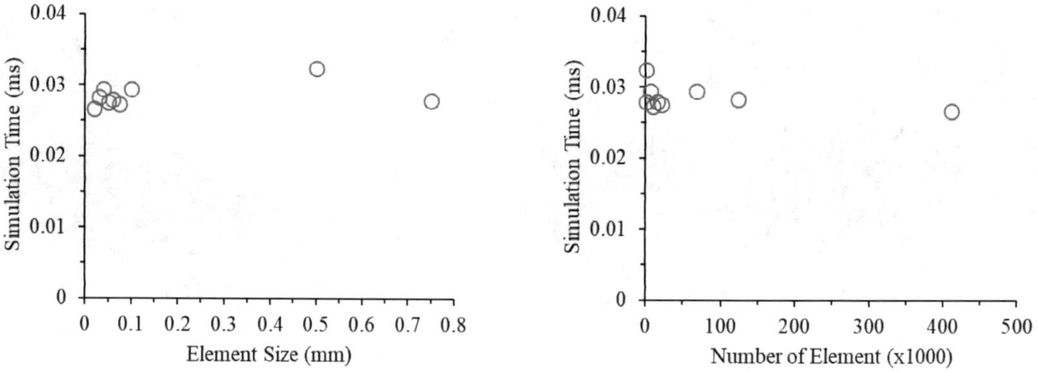

(a) (b)

Figure 5: Simulation Time against Element Size for stent expansion sensitivity analysis.

The element size of 0.075mm body sizing with 9854 number of elements was selected for further FEA simulation. This selection based on result of percentage error. The lowest percentage error gives more accurate result compared to other element size with own percentage error as shown in Table 1. Reference mesh for percentage error is highest elements which is 411942 elements and 0.02mm body sizing because high elements number claimed [14] to produce more accurate result. High mesh produces accurate result however taking more than half day of simulation time for this stent expansion simulation is very time consuming for eight speed pressure application testing to failure for stent expansion. From the sensitivity analysis in Figure 5, the mesh size is reduced from 0.02mm body sizing to 0.075mm body sizing which produce almost similar results and decrease the cost of simulation time.

3. Effect of pressure speed

The speed pressure application (SPA) was taken as loading in this simulation. There are 8 different value of speed pressure which are 400, 350, 300, 250, 200, 150, 100 and 11.3 MPa over 1ms in this paper. The value of speed pressure application was referred from Chua et al. [7] that was used only two speed pressure application for stent expansion simulation which is 400MPa/ms and 11.3MPa/ms.

Present study used higher than SPA, more element size and with sensitivity analysis to define better accurate mechanical stent structure response from different speed input variability by FEA. The stent structure undergoes stress deformation from elastic to plastic and continue until reaching material ultimate tensile stress (UTS) that considered failed [7]. The application of pressure on stent internal surface shown on Figure 6.

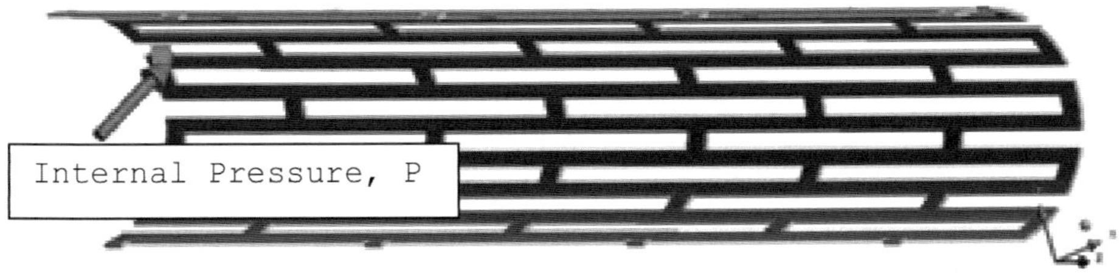

Figure 6: Pressure application to internal face of half stent

4. Results and discussion

There are several analysis of stent expansion by FEA. Output of this result can be used for future reference of stent speed pressure input or stent design optimization. The stent expands until its failure, the failure based on Ultimate Tensile Strength (UTS) 517MPa [7].

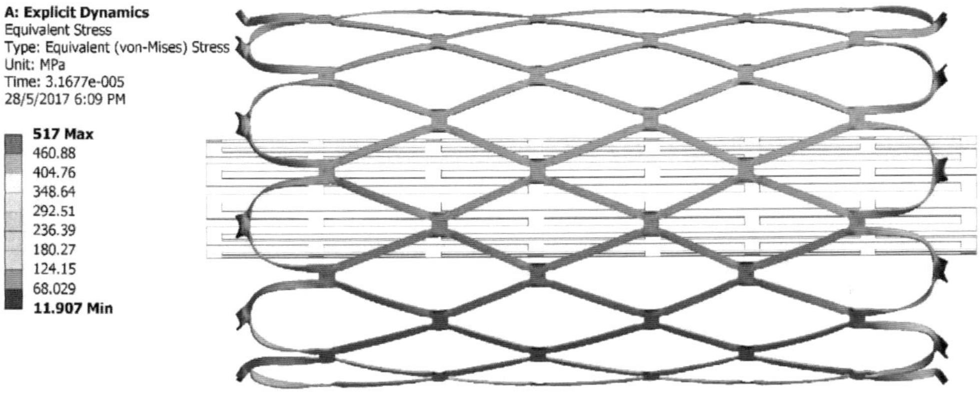

(a)

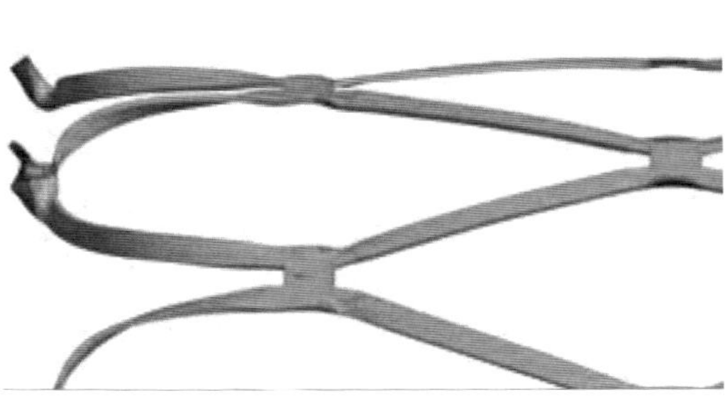

(b) (c)

Figure 7: (a) Stent expansion testing to failure 517MPa with undeformed geometry of stent, (b) Closer view of maximum stress at edge stent for 400MPa/ms SPA, and (c) Closer view of maximum stress at middle stent for 11.3MPa/ms.

Maximum Von-Mises Stress on stent structural are visualized in Figure 7(a). It is showed half of stent structure expand to failure on stent expansion process, the radial displacement and

IOP Conf. Series: Materials Science and Engineering **824** (2020) 012005 doi:10.1088/1757-899X/824/1/012005

foreshortening were analyse based on eight speed pressure application. From the figure 7(a), it can be clearly seen expansion of stent resulting increasing in diameter and decreasing in longitudinal dimension. These two output parameters referring as radial displacement and foreshortening of the stent. Stent structure consider fail after reaching ultimate tensile strength 517MPa [7]. For high SPA the stent maximum stress occurs at the edge shown by Figure 7(b) and for low SPA the maximum stress occurs at the middle shown by Figure 7(c) explaining the effect of input from low SPA which is 11.3MPa/ms producing high stress occur at four corners of the cells and low stress at the middle of the strut. Material structure erosion or fracture potentially happen at the maximum stress area. The possible crack location and fatigue fracture of stent have been reviewed by Lazim et al. [15] that mostly happen at bridge tip region similar location on this result visualization showed on Figure 7(c). Design optimization by geometrical stent structure could be focus in this area. The maximum input pressure just before failure following speed pressure application from SPA1 to SPA 8 was 10.93MPa, 9.96MPa, 8.97MPa, 7.92MPa, 6.82MPa, 5.63MPa, 4.33MPa and 1.36MPa. Result by SPA8 with 11.3MPa/ms speed pressure gives 1.36MPa maximum pressure input show 0.7% different from simulation done by Chua et al. [7] on similar speed input pressure, which is 1.35MPa of maximum pressure input. Figure 8 indicate high SPA resulting low strain and low SPA resulting high strain on nonlinear plastic region at the end of simulation, and the result is identical following SPA1 to SPA8 from high speed to low speed of pressure application on stent inner surface area.

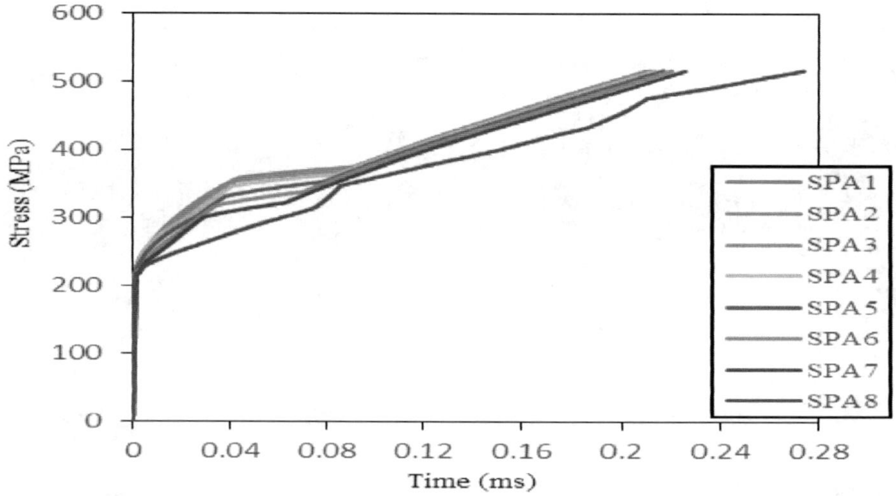

Figure 8: Stent against strain of eight speed pressure variation for stent expansion

There are two significant area defining the deformation of the stent under increasing of pressure over time, which are elastic region and plastic region. This region divides by Yield Strength of SS304 which is 207MPa. The stent behaves similar on stress and strain curve by eight variation of SPA on elastic region shown on figure 8. After the structure achieving Yield strength, under plastic region the graph indicates larger deformation from low SPA with the increasing of stress on the biomedical structure. The lowest speed pressure is curve SPA8 have the higher strain therefore indicate the largest deformation on the speed pressure setting.

IOP Conf. Series: Materials Science and Engineering **824** (2020) 012005 doi:10.1088/1757-899X/824/1/012005

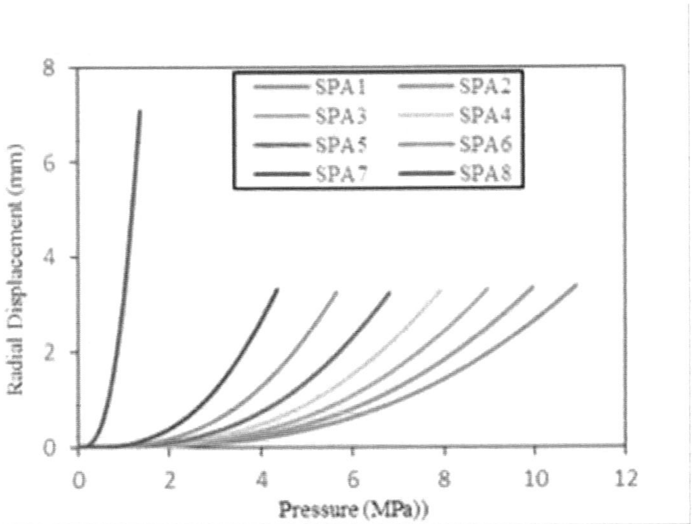

Figure 9: Radial Displacement against Pressure for 8 different speed pressure application.

Figure 9 shows the radial displacement against Pressure for 8 different speed pressure application. The lowest pressure speed denote by SPA8 has the largest radial displacement. This radial displacement data explaining on stent radial displacement selecting as surface body of stent outer surface based on polar coordinate. Data results from simulation on pressure speed SPA1 which is 400Mpa/ms gives lowest radial displacement of 3.3740mm while pressure speed SPA8 which is 11.3Mpa/ms give highest output of radial displacement of 7.0851mm. Other SPA is identical based on graph and the results are in between SPA1 and SPA8. Stent gives highest radial displacement on lower SPA. The injury on blood vessel inner surface exist on improper stent expansion process, therefore higher SPA is proposed for stent expansion based on result data. This is because low pressure speed has resulted higher radial displacement over time which can damage or rupture the artery from uncontrollable high radial displacement. The detail analysis of radial displacement is displayed on 3D view of point node and graph at end of simulation time for lowest and highest SPA illustrated in Figure 10(a) and (b). There are 9-point nodes that selected along the longitudinal axis or z-axis on polar coordinate system used in this analysis. Polar coordinate system setting z-axis for foreshortening of the stent, and x-axis for radial displacement while y-axis represent theta that indicate torsion which is not included for analysis of stent expansion in this paper. The point nodes are selected align between each other in z-axis while assuming the constant expansion in radial deformation over time. The uniform of radial displacement is investigated and showed at Figure 10(a) and (b). Uniform radial displacement is displayed by graph of SPA1 which is 400MPa/ms on figure 10(b) while non-uniform radial displacement showed by graph 11.3MPa/ms. Stent expansion require uniform radial displacement from uniform speed pressure application input to avoid inner tissue arterial surface injury.

IOP Conf. Series: Materials Science and Engineering **824** (2020) 012005 doi:10.1088/1757-899X/824/1/012005

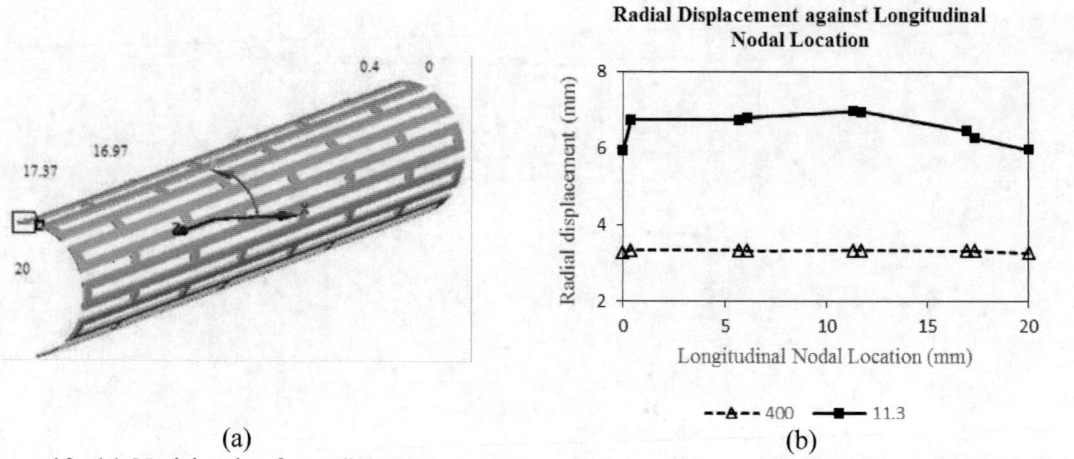

Figure 10: (a) Nodal point for radial displacement analysis on 3D stent half model, and (b) Radial displacement against longitudinal nodal location for speed pressure application of 400 and 11.3 MPa/ms.

The uniform radial displacement resulting from high speed pressure input while non uniform radial displacement resulting from low speed pressure input. Low speed pressure input has higher simulation time before structure failure and the deformation process occur differently between the middle portion of stent structure and the edge of stent structure compare to high speed pressure input. Therefore, result from the graph indicate high SPA is preferable for stent expansion application. Figure 11(a) show visualization of half stent foreshortening at the end of simulation time for 11.3MPa/ms speed pressure. It showed that large displacement on z-axis direction while the outputs of the results are controlled to give deformation only in z-axis on polar coordinate setting. The largest foreshortening showed by graph SPA8 on Figure 11(b). The simulation of stent expansion simulated by Explicit Dynamics ANSYS workbench using AUTODYN solver.

Stent foreshortening at the end of the simulation gives 4.1124mm by SPA8 and 0.8249mm by SPA1 while the other graph of SPA gives identical results in between of SPA1 and SPA8. The lower foreshortening is more favourable to assist greater stent placement inside defected blood vessel. Therefore, based the results, it indicates that larger speed of pressure application produced shorter foreshortening thus the speed pressure input by 400MPa/ms is favourable for stent expansion application.

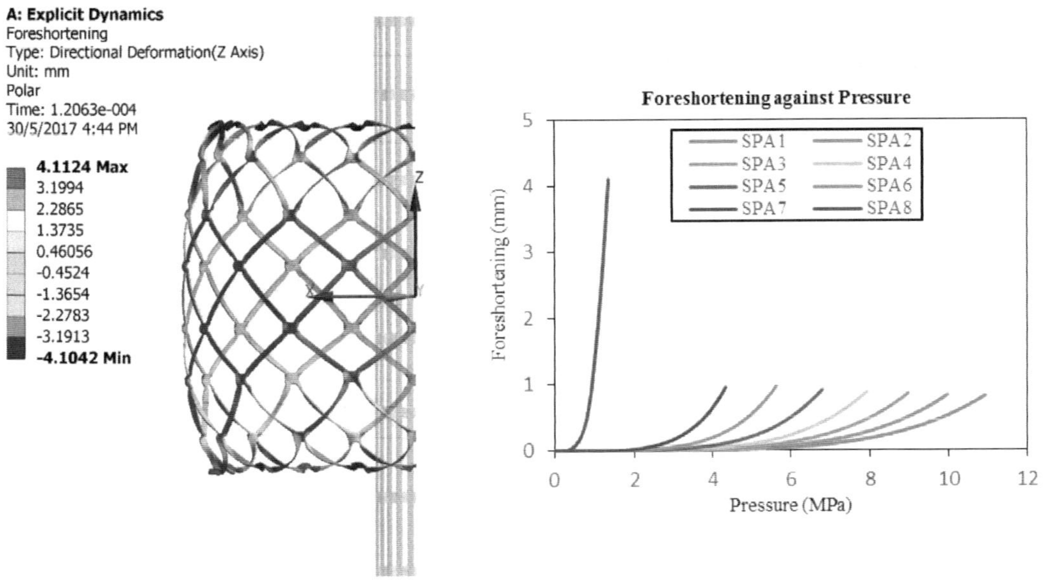

(a) (b)

Figure 11: (a) Half stent model foreshortening on expansion process for 11.3MPa/ms, and (b) Foreshortening against Pressure for eight different pressure application.

5. Conclusion

Based on the preliminary finite element analysis and investigation, there are several conclusions can be drawn:

1. Insignificant effect of different element size and number of elements on the simulation time.
2. During the expansion processes, the maximum stresses generally occurred at the joints between the structs where once the expansion reached the maximum level, the stress is beyond that the ultimate strength.

Acknowledgement

Authors acknowledge Kementerian Pendidikan Malaysia for financially sponsoring this paper through Fundamental Research Grant Scheme (FRGS) vot. 1592.

References

[1] Marple AF, Elliot EM, Hand MM, 2006, Modern Treatment for Heart Attacks Opening Blocked Arteries Quickly, *Circulation*, 114(20), 578-580.

[2] Edward H. Livingston, MD; Cassio Lynm, MA, 2012, Stents to treat coronary artery blockages, *JAMA*, **308**(17) 1824

[3] Dangas G, Kuepper F, 2002, Restenosis: Repeat Narrowing of a Coronary Artery, *Circulation*, **10**(22) 586-2587.

[4] Morrison D, Sacks A, Grover JF, Hammermeiste KE, 1995, Effectiveness of percutaneous transluminal coronary angioplasty for patients with medically refractory rest angina pectoris and high risk of adverse outcomes with coronary artery bypass grafting, *Am. J. Cardiol.*, **75** (4), 237–240.

[5] H. Li et al., 2017, Design optimization of stent and its dilatation balloon using kriging surrogate model, *Biomed. Eng. Online*,**16**, 13.

[6] Stoeckel D, Bonsignore C and Duda S, 2002, A survey of stent designs, Minim. *Invasive Ther. Allied Technol.*, **11**(4)137–147.

IOP Conf. Series: Materials Science and Engineering **824** (2020) 012005 doi:10.1088/1757-899X/824/1/012005

[7] Chua SND, Mac Donald BJ, Hashmi MSJ, 2002, Finite-element simulation of stent expansion, *J. Mater. Process. Technol.*, **20**(1–3) 335–340.
[8] He D, Liu W, Zhang T, 2015, The Development of Carotid Stent Material, *Interv. Neurol.*, **3**(2), 67–77.
[9] Hiramoto JS, Reilly LM, Schneider D, Skorobogaty BHJ, Chuter TAM, 2007, The effect of magnetic resonance imaging on stainless-steel Z-stent–based abdominal aortic prosthesis, *J. Vasc. Surg.*, **45**(3), 472–474.
[10] David Chua, Mac Donald, SN, Hashmi, MSJ, 2003, Finite element simulation of stent and balloon interaction, *J. Mater. Process. Technol.*, **143–144** (1), 591–597.
[11] McCurry MR, Evans AR, McHenry CR, 2015, The sensitivity of biological finite element models to the resolution of surface geometry: a case study of crocodilian crania, PubMed, **3**, 988.
[12] Rozakis, L. *Schaum's Quick Guide to Writing Great Research Papers*. New York: McGraw-Hill, 2007.
[13] Liu, Y, Glass, G, 2013, Effects of Mesh Density on Finite Element Analysis, *SAE Technical Paper*.
[14] Makino, M, 2008, The Performance of 10-Million Element Car Model by MPP Version of LS-DYNA on Fujitsu PRIMEPOWER, *10th International LS-DYNA Conference*, **16**, 7–10.
[15] Lazim Z, Ismail AE, Taib I, Atan BAM, 2017, A review of stent's failure on patent ductus arteriosus, *IOP Conf. Ser. Mater. Sci. Eng.*, **165** (1), 12007.
[16] Paisal MSA, Adnan SFS, Taib I, Ismail AE, Abdullah MK, Nordin N, Seri SM, Darlis N, 2017, Flow Characteristics Near to Stent Strut Configurations on Femoropopliteal Artery, *IOP Conference Series: Materials Science and Engineering,* **226** (1), 012147.
[17] Mohd Atan BA, Ismail AE, Taib I, Lazim Z, 2017, A review on fracture prevention of stent in femoropopliteal artery, *IOP Conference Series: Materials Science and Engineering*, **66** (1), 012006.
[18] Amir Paisal MS, Taib I, Ismail AE, 2017, Computational Analysis on Stent Geometries in Carotid Artery: A Review, *IOP Conference Series: Materials Science and Engineering*, **166** (1), 012003.
[19] Z. Lazim, A.E Ismail, I. Taib and B.A Mohd Atan, 2017, A review of stent's failure on patent ductus arteriosus, IOP Conference Series: Materials Science and Engineering, **166** (1), 012007.
[20] Asrizan Kaha, Ishkrizat Taib, Muhammad Sufyan Amir Paisal, Ahmad Mubarak Tajul Arifin, Al Emran Ismail, Norzelawati Asmuin, Reazul Haq Abdul Haq, Surapong Chatpun, Takahisa Yamamoto, Kahar Osman, 2018, Lumped Parameter Modelling in Femoral Popliteal Artery for Normal and Severe *Conditions, International Journal of Integrated Engineering: Special issue 2018: Mechanical Engineering*, **10** (5) 193-203.
[21] Hamdan A, Mustapha F, Ahmad KA, Mohd Rafie AS, Ishak MR, Ismail AE, 2016, The Effect of Customized Woven and Stacked Layer Orientation on Tensile and Flexural Properties of Woven Kenaf Fibre Reinforced Epoxy Composites, *International Journal of Polymer Science*, **2016**, 6514041.
[22] Ismail AE, Hassan MA, 2014, Low velocity impact on woven kenaf fiber reinforced composites, *Applied Mechanics and Materials*, **629**, 503-506.

Effect of Mechanical Properties Under Temperature influence and sucrose composition on cockleshell structure.

Ranjhini Anpalagan[1,2], Kamarul-Azhar Kamarudin[1,2], Mohamed Nasrul Mohamed Hatta[1,2], Rosniza Hussin[3] , Zaleha Mohamad[1,2]
[1]Mechanical Failure Prevention and Reliability (MPROVE) Research Centre, Crashworthiness and Collisions Research Group (COLORED),
[2]Faculty of Mechanical and Manufacturing Engineering, Universiti Tun Hussein Onn Malaysia, 86400 Batu Pahat, Johor.
[3]Faculty of Engineering Technology,Departmennt of Mechanical Engineering Technology, Universiti Tun Hussein Onn Malaysia, Pagoh Higher Education Hub, 84600 Pagoh,Muar, Johor.

E-mail: ranjianbu46@gmail.com

Abstract. Cockle shell is one of the marine organisms that easily found in Malaysia. This invertebrate type of animal lived by depending on their shell that contains of high calcium carbonate which makes it so tough and can be explore for further application. In this study, a ceramic material was fabricated from cockle shell and sucrose as a binder. The cockle shell was crushed and milled into small size at around 160 µm -200 µm. The cockle shell powder then was mixed and compacted with sucrose binder by ratio of 10, 20, and 30 wt%. The drying process was carried out in 2 methods which are heating at 150°C for one hour and normal drying at room temperature for 1 day. The effect of binder percentage and drying method on flexural properties, density and porosity of ceramic were investigated. Flexural test result shows that the samples with oven drying produced better resistance compared with samples dried at room temperature and the highest strength was obtained by sample with 20wt% sucrose dried in oven. The sample with 20wt% sucrose dried in oven also has the highest values of density. 20 wt% sucrose binder has higher ultimate strength and higher maximum force. The comparisons of sample morphology were determined by scanning electron morphology (SEM).

Keywords: Cockle shell, sucrose, binder, drying process, flexural test

1. Introduction

Presently, the recycling of waste by-products is a necessary part of waste management policies in many countries of the world. The main benefits of waste recycling are decrease in environmental pollution and conservation of natural resources (Onuaguluchi & K.Panesar, 2014). In addition, it was indicated that by reusing waste materials could also ensure waste conservation, and subsequently, decrease waste disposal in the involved sectors (Khankhaje, et al., 2017). Considering this approach, a study was conducted to investigate the mechanical properties of ceramic made of wasted cockle shell powder.

In Malaysia, cockle aquaculture areas extend about 10,383.09 hectares contributing a production of 78,024.7 tonnes in year 2010 (Department of Fisheries Malaysia, 2010).This amount will increase every year (Department of Fisheries Malaysia, 2010). Additionally, it indicated that the active and lucrative industry has resulted in a significant amount of waste shells (Boey, et al., 2011). Moreover, left untreated and dumped irresponsibly, cockle shell may produce unpleasant odour (Mustakimah Mohamed & Suzana Yusup, 2012).Othman conducted the cockle shell powders are studied in the making of concrete in order to reduce production cost of concrete and reduce waste of shells. (Othman et al., 2013) made use of cockle shells to replace cement and filler in concrete. It is concluded that higher cockle shell ash in concrete shows decline in compressive and tensile strength.

Thus far, Kamarul (2018) established small step in order to introduce cockle shell as green natural composite that can be replace the usage of the synthetic composite (Kamarula Azhar, et al., 2018). Therefore, this research to be carried out will be able for us to maximize the use of cockle shells as an environmentally friendly alternative material.

Besides that, previous study of Kamarula (2017) have found that cockle shell has capability to be one of the fundamental source to become protective material base (Kamarula Azhar et al., 2017). Moreover, a number of studies showed that the shell can improve the strength of the other ingredients when used as aggregate (Paper et al., 2012). Moreover, shell structure is a composite material that is formed naturally (Olivia, et al., 2017). Therefore, the present study focuses on the mechanical properties of cockle shell structure.

2. Materials and Methods

2.1 Preparation of cockle shell powder

Cockle shell for this study was collected from food based seafood area around Batu Pahat, Johor. Shells were initially cleaned using plain water to remove all dirt in the shell. The shells were dried completely under the sun. The shells were crushed to the size of 5 mm using a granulator machine. Then the granulated shells were transferred to Los Angeles abrasion machine to produce fine powder. After rotating the drum for 5000 rotations, the powder were removed and sieved at particle size between 160 to 200 μm.

The binder used in the study was sucrose at different percentages which were 10%, 20%, and 30 wt%. Shell powder, sucrose and water were blended together to produce a sample in accordance with specific ratio.

2.2 Specimen preparation and test setup

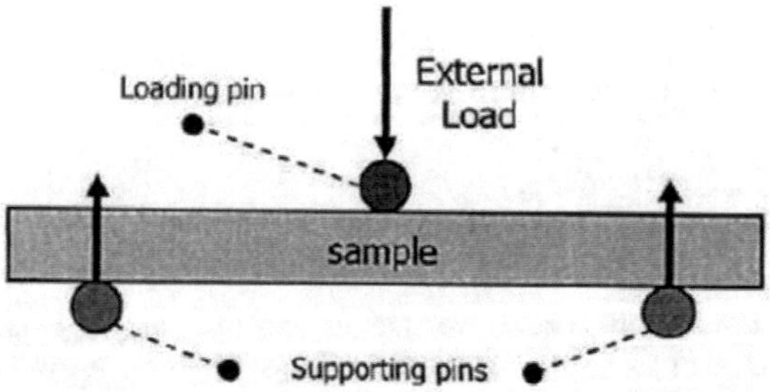

Figure 1: Schematic of flexure test of 3 point bending test (Sim, 2016)

The experiments were carried out according to the guidelines of ASTM C1161-13 (ASTM, 2013) as shown in Figure 1. The sample size used for flexural test is 10 mm wide, 7 mm thick and 100 mm length. 14g sample powder was compressed in a mould under 8 MPa pressure for 10 minutes holding time. 2 sets were prepared for each percentage of sucrose as binder. Each set has 5 samples to be tested. One set was placed in oven at 160°C for 1 hour and then kept at a normal place for 24 hours and another set was placed as normal under room temperature for 24 hours without using the oven.

3. Results and discussion

3.1 Mechanical property

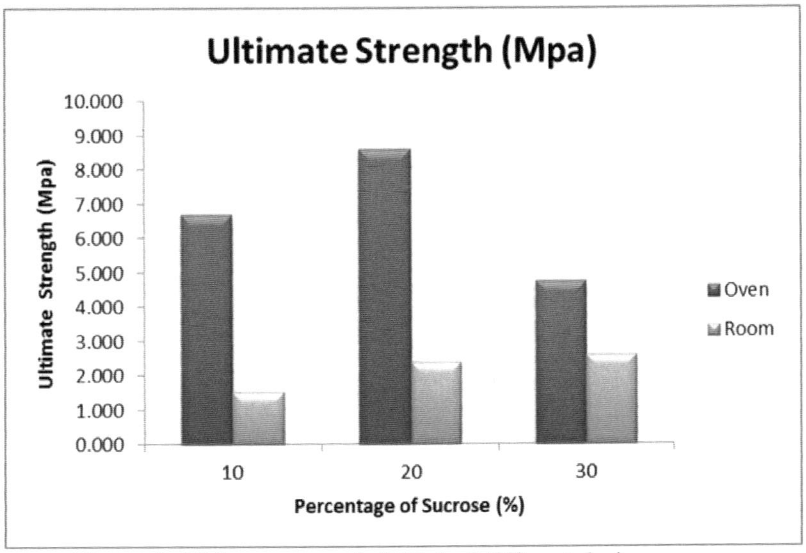

Figure 2: Ultimate strength (MPa) of different drying process

The Figure 2 shows that the oven drying process has maximum value for all three variation of sucrose binder percentages tested. Based on the graph, the specimen takes the highest strength to fracture when it was mixed with 20% sucrose in oven drying process which is 8.616 MPa. Whereas when mixed with 10wt% and 30wt% sucrose binder, its were fractured at 6.704 MPa and 4.756 MPa respectively. Therefore, it can be concluded that this happens due to the characteristics of each element in the specimen which are sucrose and cockle shell. The strongest specimen is when it only has 20% sucrose mixed in it, not more or less. This issue happened due to the 78% contain of cockles shell powder at 20% sucrose reached optimum level. If lesser cockle shell powder or higher cockle shell powder doesn't reach the highest flexural strength as 20% sucrose binder gained.

On the other hand, the strength of specimen under room temperature drying was proportionally increased with percentage of sucrose binder. However the strength of all sample was extremely lower compared to the specimens under oven drying. Even the maximum ultimate strength of room temperature 2.61 MPa is lower than the lowest ultimate strength of oven 30% sucrose binder which is 4.76 MPa. This shows that drying at room temperature doesn't react much at bonding strength between sucrose and cockle shell powder as much as react with oven drying process.

3.2 Physical property

3.2.1 Density

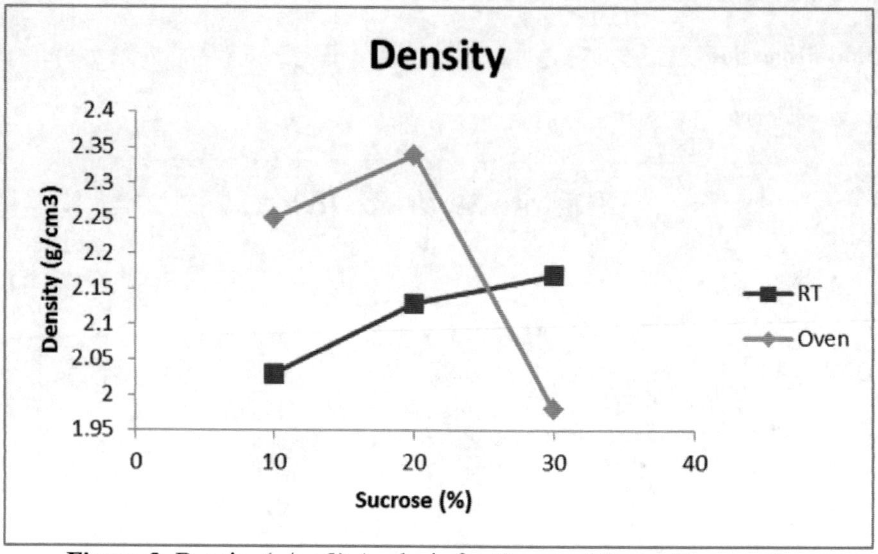

Figure 3: Density (g/cm3) Analysis for Oven and Room Temperature

The effects of drying process with different sucrose percentage on density presented in Figure 3. Density of oven drying process increased at 20% sucrose percentage as highest value but decreased at 30% sucrose as lowest value of density. It can be noticed that the trend was similar to the strength of specimen. This is indicates the density of specimen influences the strength of specimen where when the density increased, the ultimate strength also increased. Hence, the flexural strength value is depends on the density.

3.2.2 Porosity

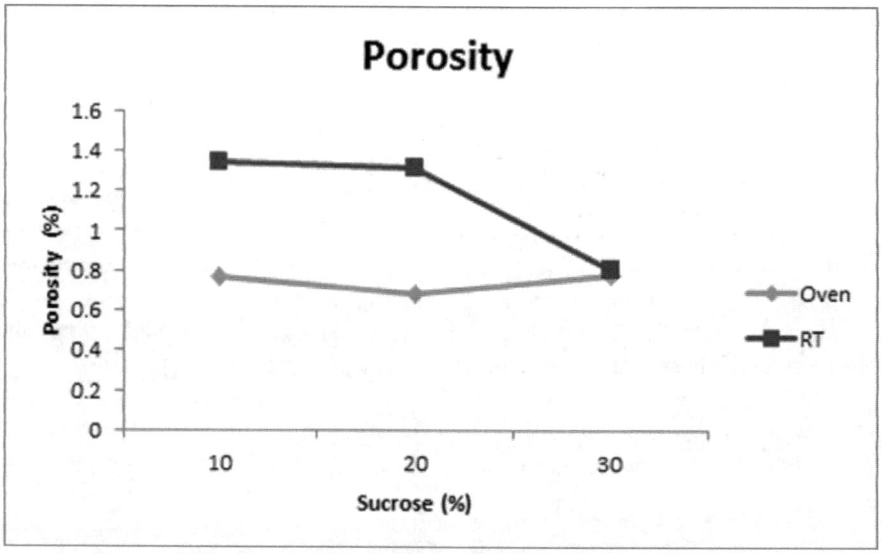

Figure 4: Porosity (%) analysis for Oven and Room Temperature

The porosity of samples from different composition and drying process are shown in Figure 4. As shown in the figure, the oven drying process has the lowest porosity which is 0.68 % and the room temperature showed the lowest porosity which is 0.81 % among the sucrose binder percentage. This is probably due to the density of samples is related could reduce porosity. Figure 4 shown the oven drying process has lowest density and room temperature has highest density. This probably, the densification of internal structure avoids the intrusion of water.

3.3 Scanning electron microscope (SEM)

3.3.1 10% Sucrose composition

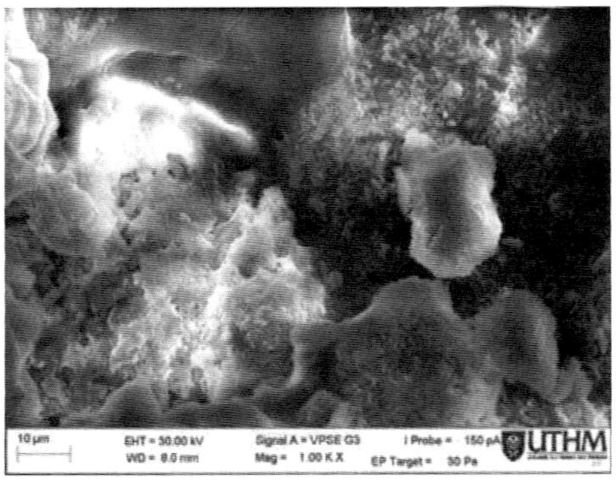

Figure 5. 10 wt% of Oven Drying

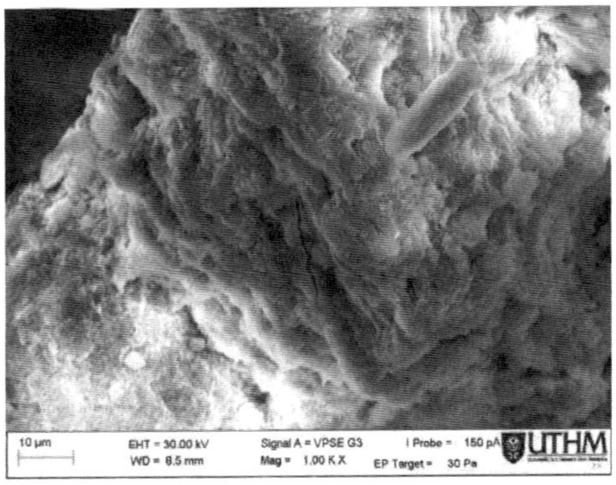

Figure 6. 10 wt% of Room temperature

Figure 5 and 6 shows the SEM images of the flexural fracture surface of 10% sucrose composition of oven drying and room temperature. The microstructure of fracture surface of sample under oven drying shows more compact than the sample under room temperature drying. The compaction of oven drying process surface has produce good strength compare to room temperature drying process.

3.3.2 20% Sucrose composition

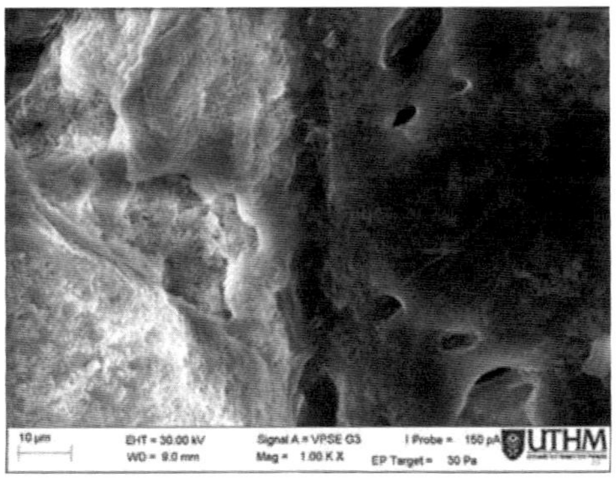

Figure 7. 20 wt% of Oven Drying

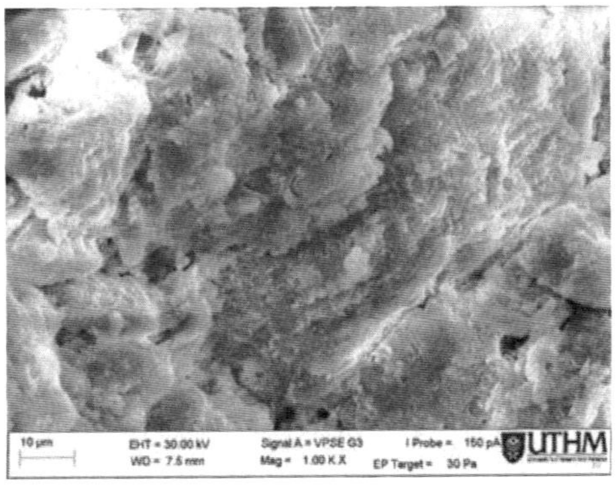

Figure 8. 20 wt% of Room temperature

Figure 7 and 8 shows the SEM images of the flexural fracture surface of 20% sucrose composition of oven and room temperature. The microstructure of fracture surface at oven shows compacted than the other percentage of fracture surface. Because of this, 20% oven has more strength than other specimens.

3.3.3 30% Sucrose composition

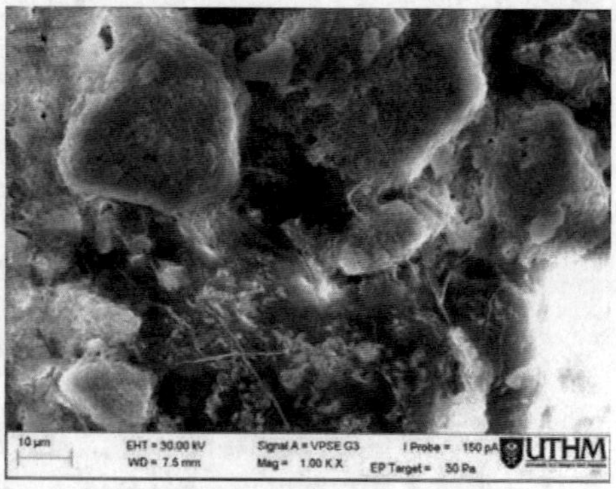

Figure 9. 30 wt% of Oven Drying

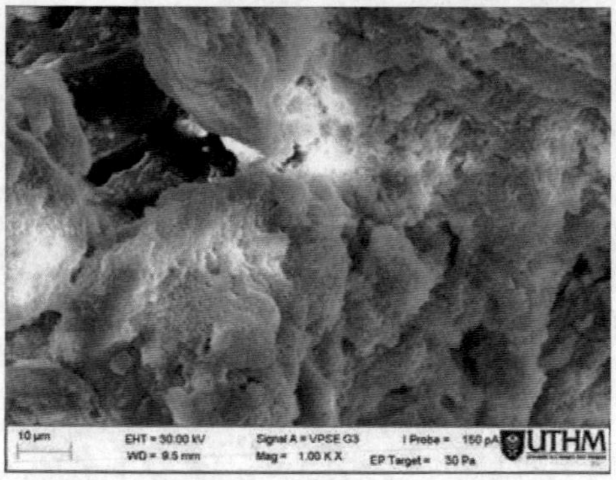

Figure 10.30wt% of Room temperature

Figure 9 and 10 shows the SEM images of the flexural fracture surface of 30% sucrose composition of oven and room temperature. The microstructure of fracture surface at oven shows compacted and less pores than the fracture surface of room temperature. Therefore, the oven drying process has good strength than the room temperature drying process.

4. Conclusion

Based on the results obtained from conducting the experiments, it is seen that the mechanical properties obtained from the test using the drying process of oven shows the best results compared to the specimens dried at room temperature. Meanwhile, when the percentage of sucrose binders of 10%, 20%, and 30% is put in comparison, the result were different for both oven and room temperature as drying process. The oven drying process shows good result compared to room temperature which shows that the specimens which had excess of moist had a weaker force compared to the less moist value which had an optimum characteristic. As a conclusion, the oven drying process is the best method for drying the specimen. Hence, the ideal percentage of sucrose mixture to increase the cockle shell specimen's strength is only 20% sucrose, 78% cockle shell powder and 2% water under oven drying process.

References

[1] Boey, P. L., Maniam, G. P., Hamid, S. A., & Ali, D. M. H. (2011). Utilization of waste cockle shell (Anadara granosa) in biodiesel production from palm olein: Optimization using response surface methodology. *Fuel*, *90*(7), 2353–2358.

[2] Kamarudin, K. A., Hatta, M. N. M., Baba, N. W. A., Hussin, R., & Ismail, A. E. (2017). Binder effect on seashell structure. *AIP Conference Proceedings*, *1891*.

[3] Kamarula Azhhar, Mohamed Nasrul Mohamed Hatta, Ranjhini Anpalagan, A. E. I. (2018). Akademia Baru Seashell Structure under Binder Influence Akademia Baru, *1*(1), 122–128.

[4] Khankhaje, E., Rafieizonooz, M., Salim, M. R., Mirza, J., & Hussin, M. W. (2017). Comparing the effects of oil palm kernel shell and cockle shell on properties of pervious concrete pavement ScienceDirect Comparing the effects of oil palm kernel shell and cockle shell on properties of pervious concrete pavement. *International Journal of Pavement Research and Technology*, *10*(September), 383–392.

[5] Olivia, M., Oktaviani, R., & Ismeddiyanto. (2017). Properties of Concrete Containing Ground Waste Cockle and Clam Seashells. *Procedia Engineering*, *171*, 658–663.

[6] Onuaguluchi, O., & K.Panesar, D. (2014). Hardened properties of concrete mixtures containing pre-coated crumb rubber and silica fume. *Journal of Cleaner Production*, *82*, 125–131.

[7] Othman, H., Hisham, B., Bakar, A., Don, M. M., Azmi, M., & Johari, M. (2013). Cockle Shell Ash Replacement for Cement and Filler in Concrete. *Malaysian Journal of Civil Engineering*, *25*(2), 201–211.

[8] Paper, C., Hazurina, N., Universiti, O., Hussein, T., Azmi, M., Johari, M., … Othman, N. H. (2012). Potential Use of Cockle (Anadara granosa) Shell Ash as Partial Cement Replacement in Concrete. *Potential Use of Cockle (Anadara Granosa) Shell Ash as Partial Cement Replacement in Concrete*, *2251*(August), 369–376.

[9] Supakorn Boonyuen, Monta Malaithong, & Apisit Prokaew. (2015). Decomposition study of calcium carbonate in shell. *Thai Journal of Science and Technology*, *4*(2), 115–122.

[10] Retrieved from Department of Fisheries Malaysia. *Department of Fisheries Malaysia*. (2011). Retrieved from <http://www.dof.gov.my/v2/

[11] Standard Test Method for Flexural Strength of Advanced Ceramics at Ambient Temperature. In ASTM, *C1161 – 13.*

[12] Sim, S. (2016, February). Mach-1-3-point or 4-point Bending test. *Biomementum*.

[13] Bellis, M. (2017, April 15). *History of Body Armor and Bullet Proof Vests*. Retrieved September 20, 2018, from Thought Co: https://www.thoughtco.com/history-of-body-armor-and-bullet-proof-vests-1991337

Acknowledgments

This project is financially supported by Ministry of Higher Education Malaysia and Universiti Tun Hussein Onn Malaysia under Fundamental Research Grant Scheme (1587).

IOP Conf. Series: Materials Science and Engineering **824** (2020) 012007 doi:10.1088/1757-899X/824/1/012007

Verification of the hydrostatic weighing system with existing gold purity instruments

Fazrul Mohd Nor[1,2]*, Abd Rahman Tamuri[2], Abd Khamim Ismail[2], A.A.M.Mazuki[1]
[1]National Metrology Institute of Malaysia, 43900 Sepang, Selangor, Malaysia
[2]Department of Physics, Faculty of Science, Universiti Teknologi Malaysia, 81310 Skudai, Johor, Malaysia

E-mail: *mfazrul@sirim.my

Abstract. Determination of gold bar purity remains one of the most challenging tasks in gold industry. Currently, the existing instruments have a limitation to determine the purity, which in turn have produced inaccurate gold purity measurement. For example, X-ray fluorescence (XRF) technique has a limitation to penetrate the gold bar and XRF was only used to determine the surface purity. Therefore, an improved non-destructive method and precise technique has been developed to determine the gold purity. Gold density measurement can verify the purity of the gold and in this study, the density of gold bar was calculated using a custom made hydrostatic weighing system (HWS). Several measurements of density were carried out for gold bar and tungsten bar. Using HWS, the density of gold bar and tungsten bar were 19.268 g/ml and 19.206 g/ml respectively. The expanded uncertainty was also evaluated and the value was 0.006 g/ml. The new developed HWS method was also verified with existing density measurement instruments where the same density results has been obtained but the existing instruments have produced bigger uncertainty of 0.2 g/ml. Thus, this study demonstrates that the developed HWS is appropriate to be used to measure the density of gold bar accurately. Furthermore, the new developed system of HWS have shown better measurement with smaller uncertainty and hence the improvement in gold bar purity measurement.

1. Introduction

Gold, silver, platinum and palladium are types of the precious metal and among these metal, gold remains the world's most precious commodities [1, 2]. For several centuries, it continues to be a secure investment for most of people. A gold with high purity is the most expensive metal therefore, its purity must be analysed through highly accurate and precise method [3]. The inspection and verification of the gold content in gold trading is also significant due to gold price is based on gold purity [4]. Pawn broking industry also need to measure the gold purity due to the requirement to pay the same value of gold to borrower by the pawn broker [5].

Presently, there are two main methods to determine the gold purity: Destructive method and Non-destructive method. Fire assay and Inductively Coupled Plasma Optical Emission Spectrometry (ICP-OES) are the techniques in destructive method [6-11]. However, these techniques are inappropriate because the pawnbroker cannot cutting or drilling the gold items that want to be leased as stated in Pawnbroker Act 1972. For non-destructive methods, the equipment such as X-ray Fluorescence (XRF), needle, chemical solution, tri-electronic and touchstone can be used to determine the gold purity without destruct the sample [12-14].

Gold density measurement can verify the purity of the gold and because of this several density instruments such as densimeter and weighing balance with solid density accessories (WB) are capable to check the purity of gold. They are also non-destructive equipment, simple to use and cheaper to buy. Most pawn shops and jewellery shops use these two instruments, however, the densimeter and WB cannot detect the presence of tungsten in gold because the tungsten density is very close to that of gold. Distilled water is used as standard liquid for the densimeter and WB. The distilled water is placed in special container and the design of liquid container for densimeter and WB is unique. One of the special design of liquid container for densimeter is shown in Figure 1. To minimise the error of measurement, the temperature of distilled water must be similar to environment temperature. The gold mass in air is measured by placing the gold sample in the top of pan whilst the sample will be placed in the bottom of the pan to weigh the mass in the distilled water. Then, The density and purity of the gold sample is obtained from the display of densimeter and weighing balance.

These two density instruments are also economical, simple and non-destructive method but produced inaccurate density with uncertainty between 0.1 g/ml to 0.2 g/ml [15]. The density of pure gold and pure tungsten are 19.30 g/ml and 19.25 g/ml respectively [16, 17]. Therefore, it will be ineffective for any measurement setup that will not produce better uncertainty with two decimal places in order to differentiate pure gold and fake tungsten gold [17-20]. The failure to accurately measure the gold density has greatly affected the gold industry [21].

This method also has the problem arising from the surface porosity of gold sample as the gold sample in the liquid needs to be removed first and then will be immersed back for the next measurement [22]. This procedure will cause inconsistency due to air bubbles trapped in the sample and it will influence the liquid temperature and thus the uncertainty of measurement.

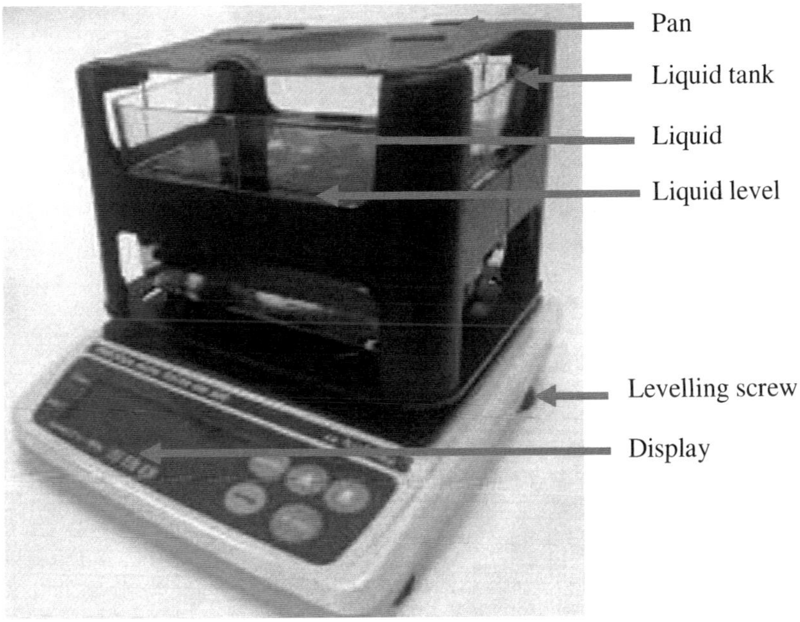

Figure 1. Composition of the densimeter

In order to get an accurate measurement, the liquid used in the densimeter and weighing balance is also must be traceable to the SI unit, where most pawnshops are not aware of this factor. The type of liquid used is also important where pure distilled water shall be used but some of the pawnshops did not replace the distilled water regularly. All these factors are important to make sure the gold purity

IOP Conf. Series: Materials Science and Engineering **824** (2020) 012007 doi:10.1088/1757-899X/824/1/012007

measurement is accurate [23-25]. Other parameters that can affects gold purity measurement are environment temperature, airflow, the liquid level in liquid container and levelling of densimeter.

Therefore, this study was conducted to explore and improve the hydrostatic weighing system by focusing on the procedure of density determination for gold purity check. The accuracy of the hydrostatic weighing system (HWS) needs to be improved at least to 0.01 g/ml due to reason that density difference between gold and tungsten is only 0.05 g/ml.

2. Method and Material

2.1. Densimeter

Densimeter is a combination of analytical balance and liquid container to measure the density of gold. Measureable range of the densimeter (AlfaMirage: Model GK-300) is 5 g to 300 g. The density readability of the densimeter is 0.1 g/ml. This method is based on Archimedes' principle where mass of the gold was determined by obtaining the different of the weight in air and the weight in distilled water. All measurements were repeated for ten times to ensure reading variability. Then using the weight difference, the density of the gold was calculated using the equation below:

$$\rho_{gold} = \frac{m_a \rho_l}{m_a - m_l} \quad (1)$$

Where:

ρ_{gold} : Density of gold (g/ml)
ρ_l : Density of liquid (g/ml)
m_a : Mass in air (g)
m_l : Mass in liquid (g)

2.2. Weighing balance with solid density (WB)

Figure 2 shows the weighing balance with density accessories consist of a support plate, glass beaker, thermometer and sample holder. The density accessories are easy to assembled and mounted to the weighing balance. The principle of this instrument is also based on Archimedes where the principle is similar to densimeter. Mass of gold sample in air and the distilled water were also measured to obtain the density of sample gold. The mass readability of the weighing balance is 0.0001 g. The density readability of the weighing balance is 0.00001 g/ml but the density accuracy is 0.1 g/ml [20].

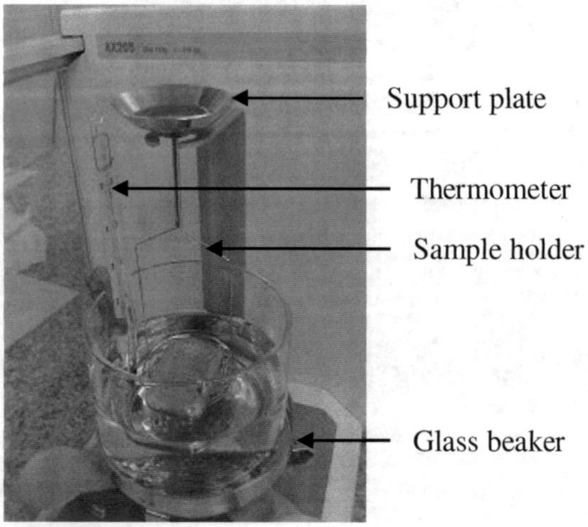

Figure 2. Weighing balance with density accessories.

2.3. Hydrostatic Weighing System (HWS)

In this study, HWS which has been set up and operates based on Archimedes' principle. In density measurement, the sample was weighed using an analytical balance in air and noted as R_1. Then, the sample was immersed in standard distilled water and weighed again as R_2. Volume of the sample, V' will change due to temperature, Δt changes and the volume thermal expansion, γ of the sample. So, volume of sample, V' is calculated as:

$$V' = \left(\frac{R_1 - R_2}{\rho_w - \rho_a} \right)(1 + \gamma \Delta t) \quad (2)$$

Finally, the density of sample, ρ_c is calculated after include the volume thermal expansion as:

$$\rho_c = \frac{R_1}{V'} \quad (3)$$

To ensure accurate calculation, all factors affecting the gold purity such as density of distilled water, ρ_w air density, ρ_a mass in liquid and liquid temperature were considered and they have been included in the equation (2). These parameters were also identified to evaluate the expanded uncertainty of measurement. Figure 3 shows the schematic diagram for HWS.

The sample was attached to the balance (Mettler Toledo; Model XP504 with capacity of 500g) using a sample holder and hooks. A digital thermometer (Model F200 with accuracy of 0.01°C) was used to measure the temperature of the distilled water in the sample bath [26]. The sample bath is a glass cylinder containing the liquid for thermal control and stability. The use of glass has enabled to view the immersion level of the sample as it is transparent. A chiller was used to stabilize the temperature of the distilled water. The air density was obtained from the equation that included air temperature, humidity and ambient pressure.

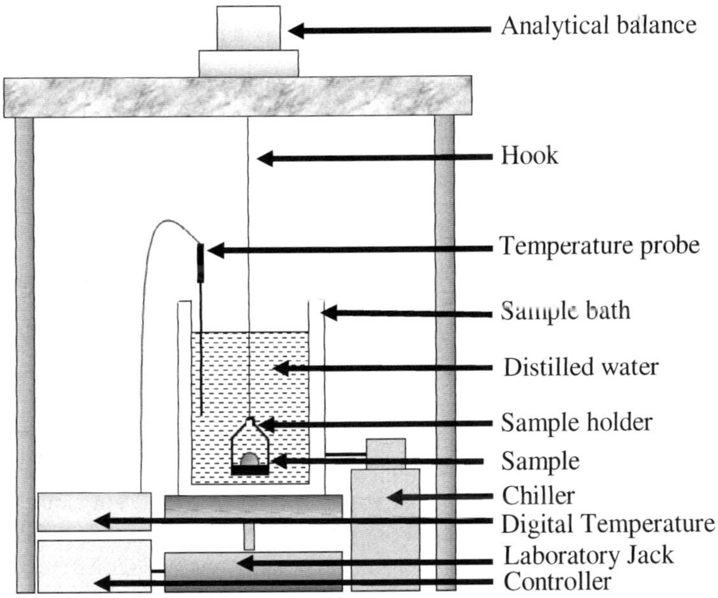

Figure 3. Schematic diagram of hydrostatic weighing system.

IOP Conf. Series: Materials Science and Engineering **824** (2020) 012007 doi:10.1088/1757-899X/824/1/012007

2.4. Material

Several samples of gold and tungsten bar with purity 99.9% were evaluated in this work due to the density of these materials is similar. So the purity and mass of gold and tungsten samples were measured using XRF (Thermo Scientific; Model ARL QUANT'X) and analytical balance (Mettler Toledo; Model XP504) respectively. Two groups of sample were measured as shown in Table 1. Sample S5 and S6 are gold samples with purity of 99.9% while S7and S8 are tungsten samples with purity of 99.9%.

Table 1. The information of the samples

Sample	Material	XRF, Purity (%)	Mass (g)	Remarks
S5	Gold	99.9	19.9990	Bar
S6	Gold	99.9	20.0113	Bar
S7	Tungsten	99.9	19.3875	Bar
S8	Tungsten	99.9	19.1537	Bar

2.5. Calculation of Performance Statistics

According to MS ISO/IEC 17043:2010 (Conformity Assessment-General Requirements for Proficiency Testing), the quality of each measurement is based on a calculated error normalised (E_n value) with respect to the stated uncertainty [27]. Hence, to ensure the measurements that were carried are reliable, E_n was calculated by comparing the result of expanded uncertainty from the uncertainty obtained by existing instruments. The calculation used is shown in the equation (4):

$$E_n = \frac{HWS - REF}{\sqrt{U_{HWS}^2} + \sqrt{U_{REF}^2}} \tag{4}$$

where:

HWS : Mean measured by HWS
REF : Mean measured by existing instruments (Densimeter and WB)
U_{HWS} : Expanded uncertainties of HWS
U_{REF} : Expanded uncertainties of existing instruments (Densimeter and WB)
$|E_n| \leq 1$: Indicates the satisfactory performance of the HWS
$|E_n| > 1$: Indicates the unsatisfactory performance of the HWS and requires root cause analysis followed by corrective actions.

3. Results and discussion

The newly developed HWS needs to be verified with other existing instruments. Verification of the HWS was done by comparing its performance with currently existing measurement instrument that are densimeter (Alfa Mirage, GK 300) and WB (Mettler Toledo, AX205). Several samples of 99.9% gold (S5, S6) and 99.9% tungsten (S7, S8) were used to verify the HWS and existing instruments. Figure 5 show the density of samples and uncertainty of measurement results for S5, S6, S7 and S8 using three different instruments. Figure 4 shows the density measurement of samples by densimeter, WB and HWS. Figure (a) shows the density of sample S5 by using HWS, densimeter and WB are 19.268 g/ml, 19.3 g/ml and 19.27 g/ml respectively. Figure (b) shows the density of sample S6 by using HWS, densimeter and WB are 19.267 g/ml, 19.2 g/ml and 19.25 g/ml respectively. Figure (c) shows the density of sample S7 by using HWS, densimeter and WB are 19.205 g/ml, 19.2 g/ml and 19.20 g/ml respectively. Lastly, Figure (d) shows the density of sample S8 by using HWS, densimeter and WB are 19.208 g/ml, 19.2 g/ml and 19.27 g/ml respectively.

The expanded uncertainty for HWS, densimeter and WB are 0.006 g/ml, 0.2 g/ml and 0.2 g/ml respectively. A list of significant measurement uncertainties for HWS obtained from Equation (2) and (3) are given in the Table 2. The expanded uncertainty determined at 95% confidence level was 0.01 g/ml.

Table 2. Analysis of expanded uncertainty of HWS at 95% confidence level

Parameter	Uncertainty (g/ml)
Mass in air (g)	0.00192
Air Density (g/ml)	0.00039
Liquid density (g/ml)	0.00001
Mass in liquid (g)	0.00214
Thermal Expansion (/°C)	0.00008
Liquid Temperature (°C)	0.00007
Combined standard uncertainty, u_c	0.0029
Expanded uncertainty, U	0.0058

From the results shown, it is clearly show that, for the same sample, the measurement obtained from HWS are consistent with the existing instruments. Using HWS method, the density of gold bar and tungsten bar were 19.268 g/ml and 19.206 g/ml respectively. This shows that, there is an improvement in new developed system of HWS that can be used to determine the purity of gold. Furthermore, the uncertainty evaluation for HWS is much smaller compared to the densimeter and WB and hence more accurate.

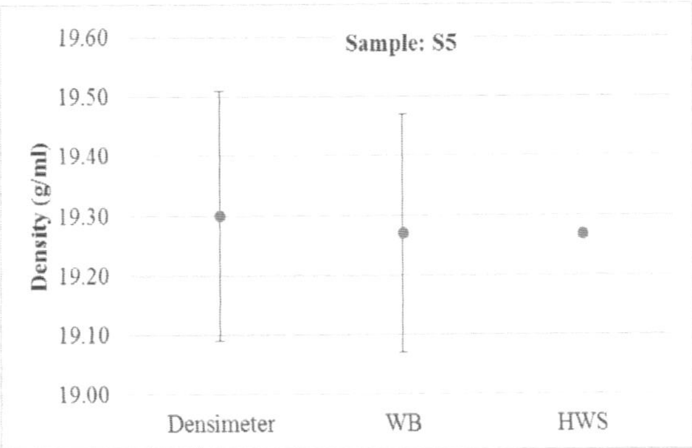

(a). Density measurement of S5 by density measurement instruments

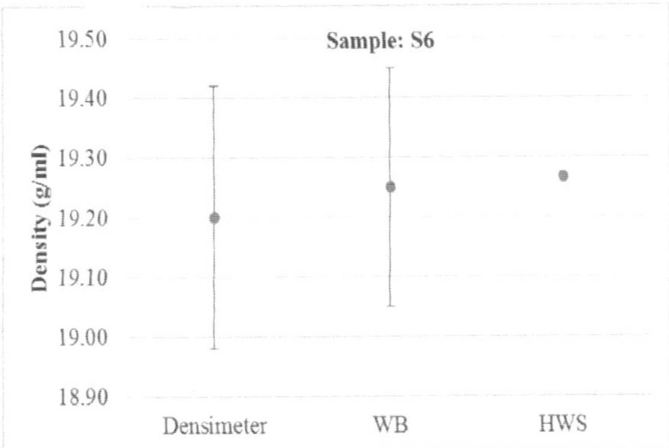

(b). Density measurement of S6 by density measurement instruments

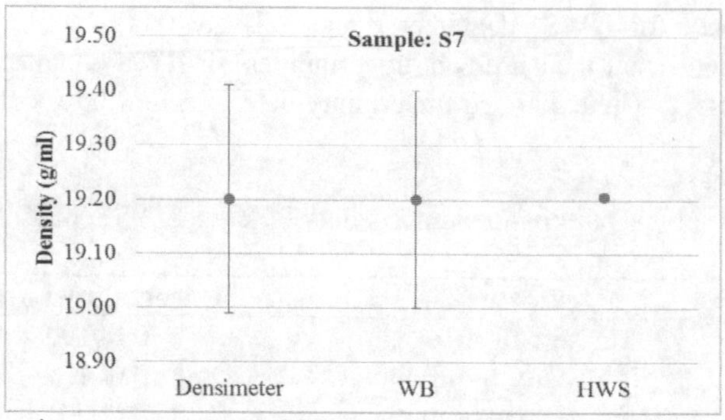

(c). Density measurement of S7 by density measurement instruments

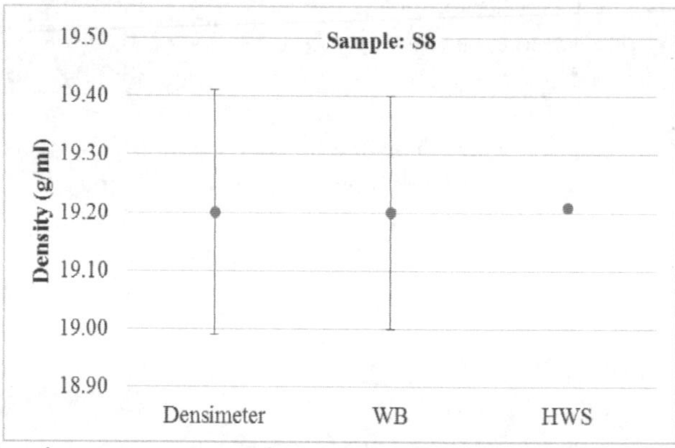

(d). Density measurement of S8 by density measurement instruments

Figure 4. Density measurement of samples by densimeter, WB and HWS.

The quality of HWS performance is based on the calculated error normalized (E_n value) with respect to the result of uncertainty as depicted in Figure 4 [28]. E_n value was calculated by comparing the result between HWS and existing instruments. Table 3 shows the E_n value of the HWS by comparing with densimeter and WB. E_n value of the HWS is below than 1, so it indicates the satisfactory performance of the HWS. Therefore, the results obtained are acceptable and thus HWS is reliable for density gold measurement.

Table 3. E_n value of the HWS by comparing with densimeter and WB

Sample	Densimeter	WB
S5	0.15	0.01
S6	0.30	0.08
S7	0.02	0.02
S8	0.04	0.04

4. Conclusions

New findings in this study has enabled HWS to be proposed as an instrument to determine the density of gold and tungsten accurately. Therefore, HWS was successfully developed to differentiate gold and tungsten and it was also verified with existing instruments. Furthermore, the quality of HWS performance is also comparable with the existing density instrument depicted by the E_n value.

5. References

1. Ding, Y., et al., *Recovery of precious metals from electronic waste and spent catalysts: A review.* Resources, Conservation and Recycling, 2019. **141**: p. 284-298
2. Cayumil, R., et al., *Concentration of precious metals during their recovery from electronic waste.* Waste management, 2016. **57**: p. 121-130
3. Jotanovic, A., et al., *Comparison of x-ray fluorescent analysis and cupellation method for determination of gold in gold jewellery alloy.* Glasnik hemicara i tehnologa Bosne i Hercegovine, 2012. **38**: p. 13-18
4. Bardi, U., et al., *Gold assay with Knudsen effusion mass spectrometry.* International Journal of Mass Spectrometry, 2008. **273**(3): p. 138-144
5. Ali, N.A.M., Y.C. Johari, and M.M.A. Fatah. *Ar-Rahnu: A Review of Literature and Future Research.* in *SHS Web of Conferences*. 2017: EDP Sciences
6. *Determination of Gold in Gold Bullion, Gold Alloys and Gold Jewellery/Artefacts - Cupellation (Fire Assay) Method.* IS 1418:2009. **Third Revision**
7. *Determination of gold in gold jewellery alloys - Cupellation method (fire assay).* International Standard, 1997. **ISO 11426**(2nd Edition:1997(E))
8. *Jewellery - Determination of precious metals in 999% gold, platinum and palladium jewellery alloys - Difference method using ICP-OES.* International Standard, 2015. **ISO 15093:2015 (E)**(2nd Edition)
9. *Copper, lead and zinc sulfide concentrates - Determination of gold and silver - Fire assay gravimetric and flame atomic absorption spectrometric method.* International Standard, 2016. **ISO 10378:2016 (E)**(3rd Edition)
10. *Methods of Test for Standard of Fineness of Precious Metals: Part 1: Gold and Articles of Gold Alloy by Cupellation Method.* MS 1365: Part 1: 1994
11. *Standard Test Methods for Determination of Gold in Bullion by Fire Assay Cupellation Analysis.* ASTM, (E1335-08)
12. Khir, M.F.A., M.B. Badri, and L. Hussain, *Critical appraisal of rahn-based Islamic microcredit facility from Shari'ah perspective.* Advances in Natural and Applied Sciences, 2013. **7**(3): p. 221-232
13. Stankiewicz, W., D. Bolibrzuch, and M. Marczak, *Gold and gold alloy reference materials for XRF analysis.* Gold Bulletin, 1998. **31**(4): p. 119-125
14. Walchli, W., *Touching precious metals.* Gold Bulletin, 1981. **14**(4): p. 154-158.
15. E.Mercer, M., *METHODS FOR DETERMINING THE GOLD CONTENT OF JEWELRY METALS.* Gems and Gemology, 1992. **28**(4): p. 222-233
16. Guo, M., et al., *Formation of scanning tracks during Selective Laser Melting (SLM) of pure tungsten powder: Morphology, geometric features and forming mechanisms.* International Journal of Refractory Metals and Hard Materials, 2019. **79**: p. 37-46
17. Abdullah, J., et al., *Non-Destructive Assaying Gold Jewellery Using Dual-Energy Micro-Computed Tomography.* Jurnal Teknologi, 2015. **73**(3)
18. Oddy, A., *Assaying in antiquity.* Gold Bulletin, 1983. **16**(2): p. 52-59
19. Raw, P., *The assaying and refining of gold, a guide for the gold jewellery producer.* World Gold Council, 1997

IOP Conf. Series: Materials Science and Engineering **824** (2020) 012007 doi:10.1088/1757-899X/824/1/012007

20. Corti, C.W., *Assaying of gold jewellery-Choice of technique.* Gold technology, 2001. **32**: p. 20-30
21. Yaacob, S.E., *CABARAN DAN ALTERNATIF PENYELESAIAN BAGI PENYEDIA PERKHIDMATAN AR-RAHNU DI MALAYSIA (CHALLENGES AND SOLUTIONS FOR AR-RAHNU SERVICE PROVIDERS IN MALAYSIA).* UMRAN-International Journal of Islamic and Civilizational Studies, 2018. **5**(1)
22. Choi, I.-M., S.-Y. Woo, and Y.-J. Lee, *New density determination method using a reference gas.* Metrologia, 2008. **45**(5): p. 524
23. Kochsiek, M. and M. Glaser, *Handbook of Metrology*. 2010: John Wiley & Sons.
24. Morris, A.S. and R. Langari, *Measurement and instrumentation: theory and application*. 2012: Academic Press
25. Jenkins, S., *Measurement and Instrumentation: Theory and Application.* Chemical Engineering, 2012. **118**(12): p. 9-10
26. Bentley, R.E., *HANDBOOK OF TEMPERATURE MEASUREMENT.: Volume 1, Temperature and Humidity Measurement*. Vol. 1. 1998: Springer Science & Business Media.
27. Yadav, S., et al., *Proficiency testing through interlaboratory comparison in the pressure range up to 70 MPa using pressure dial gauge as an artifact.* 2005
28. Bettin, H., et al., *Final report on EUROMET key comparison EUROMET. MD-K2 (EUROMET 627)" Comparison of density determinations of liquid samples".* Metrologia, 2016. **53**(1A): p. 07019

IOP Conf. Series: Materials Science and Engineering **824** (2020) 012008 doi:10.1088/1757-899X/824/1/012008

Development and Characterization of Injection Moulding Mould for Ulna Implant

N.H. Kamarudin[1], M.H.I. Ibrahim[2], S. Mahzan@Mohd Zin[2], M.H. Othman[2] , A.M. Amin[2]

[1]Faculty of Mechanical and Manufacturing Engineering, Universiti Tun Hussein Onn Malaysia, 86400 Batu Pahat, Johor, Malaysia

[2]Department of Mechanical Engineering, Faculty of Mechanical and Manufacturing Engineering, Universiti Tun Hussein Onn Malaysia, 86400 Batu Pahat, Johor, Malaysia

E-mail: nurhafizahbintikamarudin@gmail.com

Abstract. Metal injection moulding is a method by which metal parts are produced near net form by using metal feedstock. Current medical implants are products that must meet rigorous material, technology and functionality specifications. Corrosion resistance, bio-compatibility, bio-adhesion, bio-functionality and accessibility are the main medical desires of all medical implants. This work focused on the design and production of ulna plate implants using metal injection moulding. To be details on this, maximum stress and shrinkage of ulna implant components are studied. The size after demoulded from cavity was measured and compare to the cavity size. The range of shrinkage is observed by weights lost after debinding process that the result shows 0.1 gram decreased. The average maximum stress is 5.2749 MPa. Based on the result, the products are brittle after debinding process.

1. Introduction

Implant is a metal components of the body that are produced to substitute missing or damaged bones or joints. Modern medical implants are products that must meet rigorous material, technology and functionality specifications. Corrosion resistance, bio-compatibility, bio-adhesion, bio-functionality

1 nurhafizahbintikamarudin@gmail.com.

2 mdhalim@uthm.edu.my.

and accessibility are the main medical requirements of all medical implants. Ulna bone is very essential for a ideal motion in the human body. Rapid prototype method produces most human implants, but it is to costly to generate in mass production. Materials such as titanium, silicone and apatite are applied as common raw materials used to produce implant. Reduce of machining activity in producing complete products need to be considered in purpose to reduce manufacturing cost. Metal Injection Moulding (MIM) or Powder Injection Moulding (PIM) is one of the conventional manufacturing technology for fabricating this ulna implant near net shape outcomes [1]. In this study, the mold is designed to fabricate the implant components. The maximum stress of ulna implants component is measured via flexural test. Also, the correlation between the shrinkage effect of ulna implant components is investigated with cavity size. The research scope of this study is to design mold for MIM and study the correlation of injection molding material and cavity shape.

2. Literature review

Biomaterial can be defined as natural or synthetic materials that implied to support, enhance, or replace damaged tissue or biological function in human body. As known, the biomaterial engineering devices that implied in orthopaedics was called implants [2]. Purposing to meet the required characteristics which purpose to consolidate internal and external bone fracture fixation, the osteosynthesis of main implants like pins, screws, plates and nails are used with various shapes and forms [2]. The osseointegration (which referred as direct structural and functional connection between ordered, living bone and the surface of a load-carrying implant) of biomaterial is not depend only towards implant properties, but also on the host bone properties and its regenerative capabilty [3].

In United States of America, 280 000 of hip fractures, 250 000 wrist fractures and 700 000 vertebral are reported occur in each year which cost by $10 billion. Also, it was reported that more than 4 000 000 operations includes bone substitutions and bones grafting are conducted around the world every year, thus making the demanding of surgical market is crucial [4]. Secondary bone healing needs a degree of micro motion at the fracture area purposing to enhance the development of callus (tough bone tissues which grows around the end of the fractured bone during the healing process) formation [4]. Locking plates are intended to hold the fracture debris while allowing a restrained micro motion which allows a quick secondary bone healing. Whilst, the ulna-shortening osteotomy is a normal operation for multiple ulna-sided wrist disorders. Remarkably, the ulna plates are applied in ulna shortening osteotomy [5]. The implants applied in orthopaedic surgery should gratify numerous biomechanical and biocompatibily properties with referred to human body. The Ti-6Al-4V alloy able to be coated with alumina or zirconium ceramics which existing anti oxidant characteristics (bioinert ceramics), or by hydroxyapatite ceramics which have osteoconductive characteristics (bioactive ceramics). Mechanical strain should have a main effect on the interactions between biomaterials and bones [6-7]. Hydroxyapatite (HA) is known as bioactivity that commonly applied as an implant material in biomedical industries. However, its utilization is severely restrained because of the poor mechanical characteristics, which might results in discomfort of instability and reduce implant life cycle in presence of the body fluids, and local loading as compound stresses [8].

Metal injection molding (MIM) technology is become a greatly attractions among researchers purposing to produce small and intricate components in high volume mass production. Generally, there are four stages of this technology, likes mixing, molding, debinding and sintering. First, the fine metal powders are mixed with binders purposing to produce homogeneous feedstock. Next, the cool feedstock is crushed into small pellets form before undergoes the next MIM processing, which is the injection molding. This process is conducted in order to transform the molten feedstock into a required shape metal components. The binder removal is necessary to be conducted without gratify the ability of producing the free defect brown components, so called debinding process. Lastly, the brown/debound components are sintered at high temperatures often near to the theoretical.

The work on cavity and core is by far the most significant. The work on cavity and core can be categorized regarding to its form whether it is a simple or complex nature. For example, the cicular or rectangular cavity and core moulding is easier to be fabricated [9]. Runner connects the sprue via the gate with the cavity. The materials are melting and filling up throughout the cavity during moulding process at a given temperature and pressure.

Obviously, the runner dimension is depending on the maximum thickness of injected components and the type of material that will being prepared. The wider runner's cross section resulting in thicker injected components. Fundamentally, the runner cross section should be around 1 mm larger than the thickness of injected components [10]. The molten feedstock fill up the mold cavity via a sprue which is mostly machined in the sprue bushing. It is commonly being the thinnest dimension of the whole system. Then, by considering various prerequisite, the size and location of sprue is decided. It should be small as possible, thus the material is heated without damage by shear. Also, it should be easy to demould during ejecting the implant components, and permitting automate runners separation from the mould, without leaving any blemish on the injected implant components. Besides, the gate is details designed in various configurations [10-11]. The parting line generation method is proposed for a triangular subdivision of the surface model components. In this method, a draw direction is firstly selected. Remarkably, the alloy steel is applied for fabricating the mould cavity and any plates which directly contact with moulding materials. As known, the alloy steels are able to be used in both states, whether it is soft or fully hard, depend on the applications [12]. The machining operations should be consist of four geometrically defined cutters (e.g., milling, turning, drilling and sawing) and geometrically undefined cutter (grinding, honing and lapping). Then, the modern tooling machine for fabricating the mould are performed by multi axial CNC control which highly precise for positioning systems. The deviates of ideal geometrical contour of the cavity surfaces, e.g., ripples and roughness, might diminish the appearance in the particular and form "undercuts", which increase the necessary of releasing the force [10].

The binders selection is crucial to ensure the feedstock is well melting during injection molding process, which specifically for the melt front temperature difference and cooling time. According to velocity-related outputs, the powder and binders selection are crucial in order to control the maximum shear rate [13]. When shrinkage is decrease, the heating rate is increased, holding temperature is decreased and holding time is decreased. In reducing the shrinkage effects, the vacuum furnace is used to conduct debinding process, however it is useful for the length and width dimensions but not for thickness dimension [14]. The successful of the whole MIM processing is crucially depends on the rheological behavior of the feedstock. As known, the required behaviour is shear thinning, so called by pseudoplastic. For high powder loading of bimodal feedstock is more sensitive towards temperature. The flexure strength and green density of bimodal powder system are proportionally to powder loading [15]. Besides, the wear resistance behaviour of the tested materials are very depending on the feedstock homogeneity, ability of work-hardening and plastic deformation [16]. The high volume ratios (*As*/*V*) samples need less time to perform debinding process due to larger solvent accessibility to extract the binders from the implant components. Higher volume ratios, *As*/*V* cause an increase in volumetric pore frequency due to capillary pressure effect, and the temperature plays a major rule in solvent debinding by the changes of solubility and diffusion coefficient. The debinding optimization in MIM might enhance the rate production of MIM components, and minimizing defects that might be occured in subsequent processing [17].

3. Methodology

By using Solidwork software, the mold insert are drawn equal to the existing dimension, except for runner, cavity and gate system are differs. In this research work, the used material is Mild Steel AISI 1020, which also applied for, e.g., turning, milling, drilling and tapping operations, by practicing a suitable feeds, tool types and speeds. Two pieces of Mild Steel AISI 1020 plates are used in this research work, which dimension of 110 mm length x 80 mm width x 30 mm height. All the injected implant components should be milled and grind in order to obtain square shape components which accordingly to the present mould block dimensions.

In order to produce square shape of mould insert, the conventional milling machine is used. Impurities/unnecessary removal must be done for acquiring the close dimension of components, by clamping both cavity and core side to the clamping tool. The all sharp edges are milled until obtaining a chamfer which at least 5 ° angle. Next, the milling process is continued by making 45 ° angle chamfer with dimension of 3.5 mm heights, purposing to ensure that the components are feasible while fitting inside the present mould block.

Next, the components are being subjected to the process of finishing which conducted to gratify an accurate dimension of components. After completing thee surface grinding processes, both cavity and core side are machined using CNC machine, in order to produce gates, runner, cavity and core in more precise dimensions and shapes. The components (workpieces) are drawn by Solidwork software and generated to CNC machine which performed using Mastercam software. Purposing to measure an accurate outside dimension of mould insert is fabricated, it was checked by employing Coordinate Measuring Machine (CMM), before it is fabricated inside the mould block. After completing the injection moulding process, the implant components are measured and being subjected for debinding process before undertakes the flexural ISO 178 test.

4. Result and discussion

The results observe according to the geometrical size of cavity side, core side and injected components which consist of length, width and thickness. The sample was tested by flexural ISO 178 test. The shrinkage was determined by measuring the injected components weight before and after debinding process.

4.1. Measurement of Original Mould Insert

Table 1 presents the measurement of original mould insert by using CMM. The measurement is taken to produce a new mould insert specifically for fabricating the ulna plate shape, using CNC milling machine. The original mould insert measurement consists of the length, width and thickness. Since the existing mould block does not have an original drawing, the measurement must be taken to ensure the new mould insert fit inside the mould block without any gaps which may resulting the damage towards the mould block and mould insert itself.

Table 1. The original mould insert dimensions

Type	Length (mm)	Width (mm)	Thickness (mm)
Cavity	99.9549	69.9851	25.0356
Core	99.9231	69.9878	15.0300

4.2. Fabrication of Core Side

Purposing to fabricate the core side, the CNC milling machine and surface grinder machine are used. The machining process consist of 4 tapping holes and 1 sprue bushing hole. In order to measure the core side, the CMM is employed purposing to ensure the precise dimension obtained is approximately same to the original core size. Table 2 tabulates the core side dimensions after machining process. Precise core side dimension is crucial in order to ensure the mould insert is fitted inside the mould block without any gap. The thickness between original core and fabrication core not equal because of there are several parts that need to fit inside the cavity.

Table 2. Core side Measurement

Length (mm)	Width (mm)	Thickness (mm)	Hole Diameter (mm)
99.5179	69.9655	18.1884	12.0311

4.3. Fabrication of Cavity Side using Ejector Pin System

In order to fabricate the cavity side mould, it was done by CNC milling machine, surface grinder and drilling. As known, the ejector pin holes are used for allowing the ejector pins for pushing out the injected implant components after the filling phase is completed. There are 5 ejector pins, which consist of 4 pins for ejecting the implant component and the other one pin is for ejecting the sprue runner. Table 3 tabulates the measurement of cavity side mould after the machining process. The mould insert is divided into two part, e.g., cavity side and core side. Then, the cavity side is consist of 5 ejectors pin hole, which made up of 4 holes for ejecting part and 1 for ejecting the sprue runner. Subsequently, the all dimensions are being compared. But, there are some errors regarding to the original mould insert and the fabricated mould insert. However, it was in acceptable range because the fabricated mould insert was still in the tolerances range of +/-0.5 mm.

Table 3. Cavity side measurement

Specification		Dimensions (mm)			
Length (mm)		99.8915			
Width (mm)		69.9855			
Thickness (mm)		25.1884			
Ejector hole Ø, (mm)	**Part**	1	2	3	4
		6.0112	6.0121	6.0211	6.0111
	Sprue runner	5.0311			

4.4. Result of Implant Components

The complete injected implant components are measured purposing to compare the cavity size of the mould insert, in regards to study the shrinkage effect. All injected implant components are placed on granite table and they are measured using digimatic calliper. It was taking 5 repetitions in measuring the shrinkage effect of the injected implant components with the mould cavity, and it was presented in Table 4. In this research work, hydroxyapatite, palm stearin and Low Density Polyethylene (LDPE) are chosen to be mixed and forming a homogeneous feedstock. Whereas, Table 5 presents the injection molding parameters which conducted by employing NP7 Real Mini of horizontal plastic injection moulding machine, Polymer and Ceramics Laboratory, UTHM.

Table 4. Products dimensions

Samples	**Length (mm)**	**Width (mm)**	**Thickness (mm)**
Sample A	39.88	9.85	2.89
Sample B	39.86	9.88	2.85
Sample C	39.99	9.79	2.91
Sample D	39.87	9.91	2.87
Sample E	39.79	9.89	2.93

Table 5. Injection parameters

Type		**Machine setting**
Heating Temperature (°C)	Nozzle	165
	Front	165
	Medium	165
	Rear 1	155
	Rear 2	150
Pressure (P), MPa		1.1
Injection Velocity (%)		25
Injection Time (s)		3

4.5. Solvent Extraction

The weight loss of the palm stearin is measured and tabulates in Table 6. The injected implant components are soaked in heptane solution within 5 hours at 70 °C. The samples are weighed for each 1 hour. It can be seen that the all samples weight are decreased with regards to immerse time.It was found that weight loss recorded was in range of 0.09 g to 0.10 g. After the solvent extraction process, the injected implant components become more brittle and needs appropriate ways in handling it for the next subsequent processing.

Table 6. Products dimensions

Samples	**Weight before debinding (g)**	**Weight after 1 hours (g)**	**Weight after 3 hours (g)**	**Weight after 5 hours (g)**	**Average weight lost (g)**
Sample A	1.182	1.164	1.128	1.092	0.09
Sample B	1.172	1.154	1.118	1.082	0.09
Sample C	1.201	1.185	1.149	1.111	0.09
Sample D	1.191	1.167	1.129	1.091	0.1
Sample E	1.210	1.194	1.162	1.130	0.08

4.6. Flexural Test

The flexural strength is defined by the components ability to withstand the deformation under the given load. The flexural test is conducted in order to measure the green strength of the implant components. The 5 repetitions is made in order to obtain an accurate green strength data. Three types of data are recorded for this flexural test, i.e., maximum stress, maximum force and maximum displacement.

Table 7. Flexural test results

Samples	**Maximum force (N)**	**Maximum stress (MPa)**	**Maximum displacement (mm)**
Sample A	37.4812	4.6852	0.6470
Sample B	44.7000	5.5875	0.7720
Sample C	38.9719	4.8715	0.6780
Sample D	43.0688	5.3836	0.7880
Sample E	46.7750	5.8469	0.6590
Average	42.1994	5.2749	0.7088

5. Conclusion

This research work aims to design and fabricate the mould insert for metal injection moulding, in producing the ulna plate implant components. The maximum stress of ulna implant components are studied via flexural test and the shrinkage effect of ulna plate implant components are studied, which correlates to the cavity size. Since this is a preliminary research towards implants material, the material which compatibles to human bone are not covered in this research work, but only covered the design process, e.g., ulna implant component geometries after demoulded and debinding, and correlation between injection moulding material and cavity shape.

IOP Conf. Series: Materials Science and Engineering **824** (2020) 012008 doi:10.1088/1757-899X/824/1/012008

6. References

[1] Shimizu T 2001 Production of large size parts by MIM process Journal of Materials Processing Technology 119 199-202

[2] Rodríguez-González F A 2009 Introduction to Biomaterials in Orthopaedic Surgery (United States: ASM International)

[3] Fini M, Giavaresi G, Torricelli P, Borsari V, Giardino R, Nicolini A and Carpi A 2004 Osteoporosis and biomaterial osteointegration Biomedicine & Pharmacotherapy 58(9) 487-493

[4] Brydone A S, Meek D and Maclaine S 2010 Bone grafting, orthopaedic biomaterials, and the clinical need for bone engineering Proc. of the Institution of Mechanical Engineers Part H: Journal of Engineering in Medicine 224(12) 1329–1343

[5] Larson A N, Rizzo M 2007 Locking plate technology and its applications in upper extremity fracture care Hand Clinics 23(2) 269-278

[6] Palma F D, Douet M, Boachon C, Guignandon A, Peyroche S, Forest B, Alexandre C, Chamson A and Rattner A 2003 Physiological strains induce differentiation in human osteoblasts cultured on orthopaedic biomaterial Biomaterials 24(18) 3139-3151

[7] Merhar J R 1990 Overview of Metal Injection Moulding Metal Powder Report 45(5) 339-342

[8] Li H, Ka K and Cheang P 2002 Impact formation and microstructure characterization of thermal sprayed hydroxyapatite/titania composite coatings Biomaterials 24(6) 949-57

[9] Pye R. G W 1986 Injection Mould Design (Berlin: Springer)

[10] Menges G, Michaeli W and Mohren P 2001 How to Make Injection Molds (Munich: Hanser)

[11] Kwong K W, Ming Fai M, Sze W S and Sooithor T 1988 A method for generation of parting surface for injection moulds *Proc. of 4th Int. Conf. on Computer aided Production Engineering* (University of Edinburgh) p 401-408

[12] Jones P 2008 The Mould Design Guide (United Kingdom: Smithers Rapra)

[13] Ahn S, Seong J P, Lee S, Atre S V and German R M 2009 Effect of powders and binders on material properties and molding parameters in iron and stainless steel powder injection molding process Powder Technology 193(2) 162-169

[14] Loh N H and German R M 1995 Statistical analysis of shrinkage variation for powder injection molding Journal of Materials Processing Technology 59(3) 278-284

[15] Jamaludin K R, Muhammad N, Ahmad S, Ibrahim M H I, Mohd Nor N H and Daud Y, 2011 Injection moulding temperature and powder loading influence to the metal injection moulding (MIM) green compact Scientific Research and Essays 6(21) 4532-4538

[16] Schneider J, Iwanek H and Zum Gahr K H 2005 Wear behaviour of mould inserts used in micro powder injection moulding of ceramics and metals Wear 259 1290-1298

IOP Conf. Series: Materials Science and Engineering **824** (2020) 012008 doi:10.1088/1757-899X/824/1/012008

[17] Oliveira R V B, Soldi V, Fredel M C and Pires A T N 2004 Ceramic injection moulding: influence of specimen dimensions and temperature on solvent debinding kinetics Journal of Materials Processing Technology 160(2) 213-220

Acknowledgments

The authors of this research work are gratefully gratitude for their financial support and guidance by Geran Penyelidikan Pascasiswazah (GPPS), Pusat Pengurusan Penyelidikann (RMC), Universiti Tun Hussein Onn Malaysia (UTHM).

IOP Conf. Series: Materials Science and Engineering **824** (2020) 012009 doi:10.1088/1757-899X/824/1/012009

Investigation on the Performance of a Novel Dehumidifier for a Combined Solar Chimney and Liquid Desiccant Plant

C Y Ang, A J Alimin and S K Mahal
Centre for Energy and Industrial Environment Studies (CEIES), Faculty of Mechanical and Manufacturing Engineering, Universiti Tun Hussein Onn Malaysia, 86400 Parit Raja, Batu Pahat, Johor, Malaysia

E-mail: angcy95@gmail.com, ajais@uthm.edu.my

Abstract. A dehumidifier was one of the main component in the hybrid liquid desiccant system and solar chimney micro-plant. The performance of the hybrid liquid desiccant system and solar chimney micro-plant was strongly depended on the performance of it dehumidifier, but it is yet to be analyzed in any of the previous studies. Therefore, moisture removal rate (MRR) and dehumidification effectiveness (ε) were two parameters that had been chosen to evaluate the performance of the cross flow, cellulose panel structured bed dehumidifier in this study. Calcium Chloride ($CaCl_2$) desiccant solution was the only liquid desiccant solution used in this application. From the experiment, it was found that the highest effectiveness of the dehumidifier was about 22.78% and the highest moisture removal rate was about 0.22ml/s. Besides that, parametric analysis had been conducted on five inlet parameters: desiccant inlet concentration, desiccant inlet temperature, inlet air humidity ratio, inlet air mass flow rate and inlet air temperature; the most influencing parameter was found to be the inlet air humidity ratio, y_{a1}. MRR increased by 52.68% and ε of the dehumidifier increased by 48.59% when y_{a1} increased from 0.0175-0.0200. Finally, three improvement strategies had been recommended: increasing the concentration of liquid desiccant, changing the type of liquid desiccant and increasing the thickness of the dehumidifier.

Keywords: solar chimney, fresh water, generator, water security, solar energy

1. Introduction

In human daily life, water and energy are two indispensable item for us to continue our living[1], as an example it is a natural resource that humans need in order to maintain the function of organ and regulate body temperature, while energy are needed for light. As the human population kept increasing globally, the dependence of fossils fuels and the scarcity of other non-renewable resources has force the scientist and researchers worldwide to find out alternative strategies for generating fresh water and electricity to fulfill the growing demand. Therefore, a technology of hybrid liquid desiccant and solar chimney system micro-plant has been applied in current study. This micro plant as shown in figure 1.1 is a combination of two green technologies which are liquid desiccant system and solar chimney, also known as solar updraft tower. The earliest application of solar chimney application is on 1982, the Manzanares power plant in Spain[2]–[4]. Since then, solar chimney or solar updraft tower has caught the attention of scientist and researchers worldwide and a research campaign on this new technology has been started. A review on the concept, history and operating features of the solar chimney application has been done. Besides that, the potential in combined solar chimney with liquid desiccant system for simultaneous production of electricity and fresh water through atmospheric humidity has also been discussed by the author.[5]

Figure 1.1. The hybrid liquid desiccant system and solar chimney micro-plant

1.1. Literature review

As mentioned in the tittle, the investigation on the performance of a novel dehumidifier for a combined solar chimney and liquid desiccant plant, the main focus of this study is the novel dehumidifier used in this system. As a result, table 1.1 has list out all the similar previous studies related to the performance of liquid desiccant dehumidification system (dehumidifier). In this investigation, the flow configuration is cross-flow for the dehumidification systems of liquid desiccant apply in this combined micro plant. Therefore, heat and mass transfer process for cross-flow configuration and type of desiccant solution used has been study in order to formulate all the require equation for thermodynamic analysis. All the paper listed in the table is the main references for completing this literature review.

Table 1.1. Similar previous studies on the performance of liquid desiccant dehumidification system

Author And Year	Findings	Desiccant Solution Used
Bassuoni (2014)[6]	A simple analytical method to estimate all exit parameters	$CaCl_2$
Gao (2012)[7]	Experiment investigation of heat and mass transfer	LiCl
Pineda & Diaz (2011)[8]	Performance of a small scale prototype	$CaCl_2$
Babakhani (2010)[9]	Coupled mass and heat transfer process	LiBr
Moon (2008)[10]	New mass transfer performance data	$CaCl_2$
Liu (2004)[11]	Experimental study on mass transfer performance of cross flow dehumidifier using liquid desiccant	LiBr

IOP Conf. Series: Materials Science and Engineering **824** (2020) 012009 doi:10.1088/1757-899X/824/1/012009

2. Methodology

2.1. System Description

Solar chimney usually consists of a circular greenhouse type collector and a tall tower at its center. This system application works when air flow radially inwards under the air collector and is heated by the collector floor and roof, and the hot air flow through a turbine to enters the chimney. The solar chimney in current study is applied and combined with the dehumidifier of the liquid desiccant system which generates water from the atmospheric humidity. As illustrated in figure 2.1, a typical schematic diagram of a technology for fresh water production from atmospheric humidity by using liquid desiccant system[12]. The main components in such system consist of dehumidifier, regenerator, heat exchanger, desiccant cooler, desiccant heater and also the circulating pumps. The application of solar chimney in current studies helps drawing in the wet atmospheric air from the surrounding into the dehumidifier of liquid desiccant system. As the wet atmospheric air flows into the dehumidifier, the liquid desiccant that flow through the top of the dehumidifier absorb the air moisture from the wet atmospheric air and bring the weak liquid desiccant solution (with air moisture) to the regenerator tank to produce fresh water. This kind of system usually applies near sea side, as the sea water can act as a natural sea water reservoir for cooling purpose. However, in current study this sea water reservoir has been replaced with a cooling water tank. The main focus of this study is about the cross flow, cellulose panel structured bed dehumidifier which surrounded the solar chimney and covered by transparent polyethylene sheet as shown in figure 2.2. The dimension of this dehumidifier used is 0.6m in height, 0.6m in width and 0.25m thick. The exact location of this micro plant is besides block G8 and behind the Tun Fatimah Residential Colleague of Universiti Tun Hussein Onn Malaysia (UTHM).

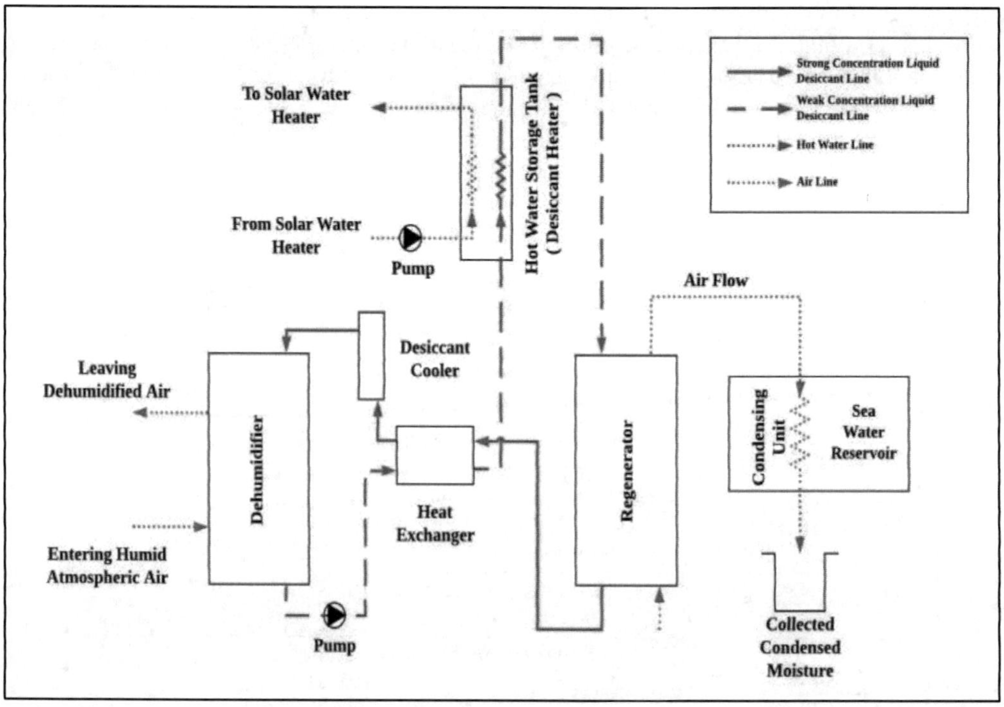

Figure 2.1. Schematic diagram of a fresh water production technology from atmospheric humidity by liquid desiccant system.[12]

IOP Conf. Series: Materials Science and Engineering **824** (2020) 012009 doi:10.1088/1757-899X/824/1/012009

Figure 2.2. Cellulose panel structured bed dehumidifier used in the combined solar chimney and liquid desiccant plant

2.2. Data Collection

To obtain all the operating parameter of the combined solar chimney and liquid desiccant micro plant in current study, a few instruments are needed as shown in table 2.1. The step and method to collect the parameters of solar chimney are shown in figure 2.3. A ladder will be used to climb over and the probe will be inserted to the chimney to obtain all the required operating parameters. A cloth will be put around the probe during measurement to prevent the hot air from flowing out from the cap. Air temperature, relative humidity and air velocity in the chimney are being measured at the designated cap of the solar chimney by using TSI VELOCICALC Plus Multi-Parameter Meters.

Table 2.1. Similar previous studies on the performance of liquid desiccant dehumidification system

Operating parameters	Devices
Ambient air temperature and relative humidity (Surroundings)	BENETECH Digital Multi-purpose Anemometer
Air Velocity, Air temperature and relative humidity in the chimney	TSI VELOCICALC Plus Multi-Parameter Meters
Concentration of Liquid Desiccant	Measuring Tube and Pocket Digital Scale
Inlet Liquid Desiccant Temperature	AMS V-line Thermometer
Desiccant Flow Rate	TT10 In-Line Digital Turbine Meter

Figure 2.3. Steps in collecting the data from the solar chimney

IOP Conf. Series: Materials Science and Engineering **824** (2020) 012009 doi:10.1088/1757-899X/824/1/012009

2.3. Thermodynamic Analysis

Thermodynamic analysis has been apply to this study as this system involves mass flow across the boundaries especially on the dehumidifier. Based on the conservation of mass principle, the net mass transfer, $m_{in} - m_{out}$ is equal to the net change in the system, Δm_{system} during the process. Equation (1) shows the mass balance equation which is applicable to any kind of system with any kind of process. The conservation of mass equation can be referred as continuity equation in fluid mechanics.

$$m_{in} - m_{out} = \Delta m_{system} \quad (1)$$

2.4. Moisture Removal Rate

The moisture removal rate (MRR) can be define as equation (2), where m_{cond}, is the rate of water vapor condensed by the dehumidifier kg/s.

$$m_{cond} = MRR = m_a(\Delta_y) \quad (2)$$

2.5. Dehumidification Effectiveness

Dehumidification effectiveness is one of the common performance measures for investigating the dehumidifier potential to dehumidified air. From the research paper of Moon et al.[10], equation (3) an empirical correlation of the humidity effectiveness ε which is valid for cross-flow liquid desiccant dehumidifiers have been developed.

$$\varepsilon = \frac{y_{a1} - y_{a2}}{y_{a1} - y_{eq}} \quad (3)$$

Where, y_{eq} the humidity ratio of air in equilibrium with desiccant ($CaCl_2$) solution at the interfacial area is calculated as shown in equation (4):

$$y_{eq} = \frac{0.622 p_{vap}}{1.013 \times 10^5 - p_{vap}} \quad (4)$$

As p_{vap}, partial vapor pressure on the desiccant solution surface in Pa. By using equation (5) the correlations introduced by Gad et al.[13], the partial vapor pressure on $CaCl_2$ will be calculated with this formula below

$$ln(p_{vap}) = (a_0 + a_1 X) - \frac{(b_0 + b_1 X)}{T_s + C} \quad (5)$$

Where:

- a_0, a_1, b_0, b_1 *and C are all constant value*
- X desiccant ($CaCl_2$) solution concentration, kg_d/kg_s
- T_s desiccant ($CaCl_2$) temperature in °C

IOP Conf. Series: Materials Science and Engineering **824** (2020) 012009 doi:10.1088/1757-899X/824/1/012009

3. Results And Discussion

3.1. Experimental Results

Five testing days in between 19th October 2018 to 4th November 2018 has been choose to conduct experiment study on the hybrid liquid desiccant system and solar chimney micro plant, from 9 am to 6 pm on a full day basis. All the results obtained have been analyzed in the next section.

3.2. Performance Parameters

As illustrated in figure 3.1 below, an hourly average line has been plotted and only 3 days have exceeded the average line during the peak hour at 1 pm. The highest MRR obtain is on the 28 October. Based on the data obtained, 28 October is one of the humid day experienced during the testing as most of the ambient humidity ratio is above 0.02. Another factor that leads to high MRR believes to be the ambient temperature on 28 October, an ambient temperature as high as 36°C is registered at the noon (about 1pm). On the other hand, the hourly effectiveness of the dehumidifier also give some similar results to the MRR, as there are only 3 days that exceed the average line at the peak time as shown in figure 3.2. These 3 days that exceed the average line of MRR and hourly effectiveness of dehumidifier are the same which are 24 Oct, 28 Oct and also 4 Nov. The highest effectiveness of 22.78% has been registered on the 24 October. Referring to the data recorded, 24 October is assumed to be a humid day because the humidity ratio registered at 1 pm is about 0.0221, which was the highest among the five testing days. Besides humidity ratio, the difference in humidity ratio, Δy during 24 October is 0.005, which is the highest when compared to all the data registered among five testing days. This explains why the dehumidification effectiveness is the highest during 24 October. As a results, good MRR will need a combination of high outdoor humidity ratio and high air mass flow rate, while good effectiveness need a combination of high outdoor humidity ratio and a good difference in humidity ratio, Δy.

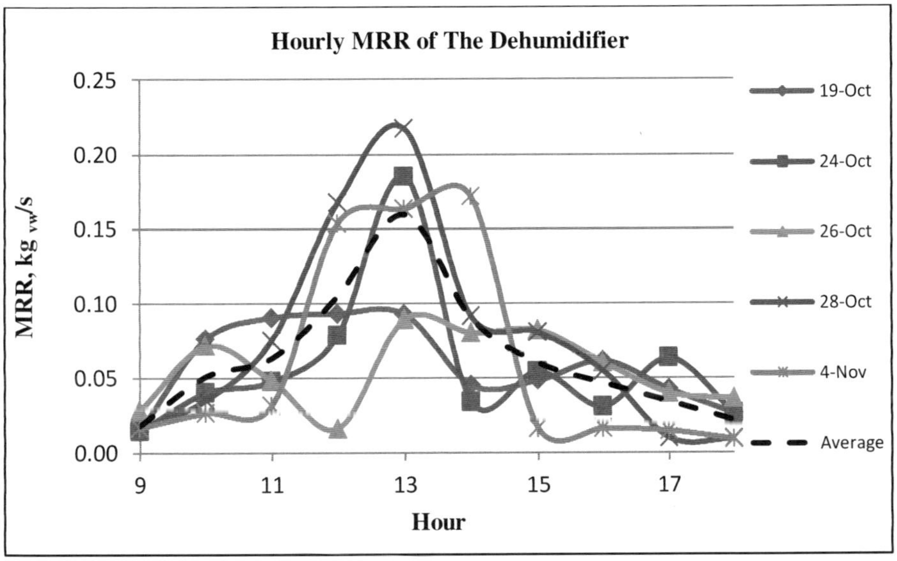

Figure 3.1. Hourly Moisture Removal Rate, MRR of the dehumidifier

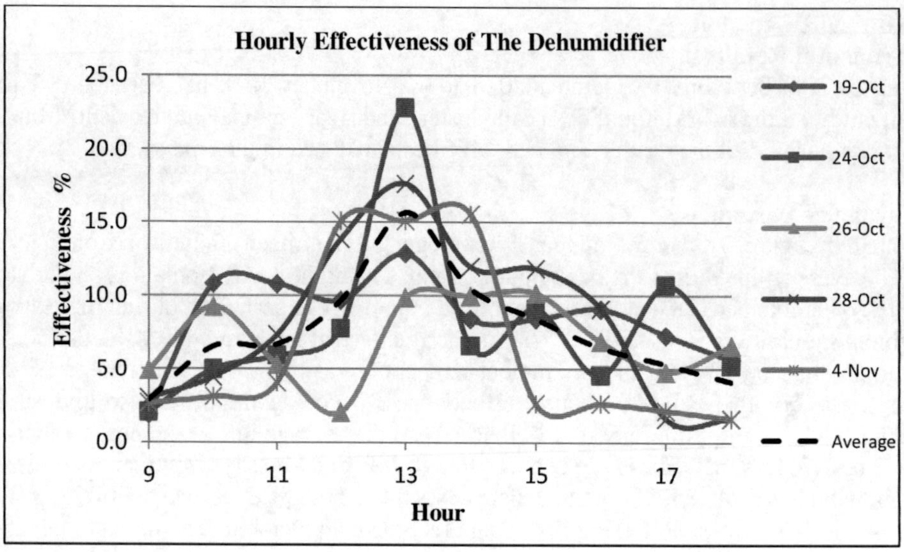

Figure 3.2. Hourly Effectiveness, ε of the dehumidifier

3.3. Parametric Analysis

There are five inlet parameters which are desiccant inlet concentration, desiccant inlet temperature, air inlet humidity ratio, air inlet mass flow rate and air inlet temperature in this study which will surely affect the performance, moisture removal rate (MRR) and effectiveness (ε) of the novel dehumidifier of a combined solar chimney and liquid desiccant plant. Parametric analysis has been conducted with the operating conditions as table 3.1 below to determine the most influencing parameters for these systems.

Table 3.1. Operating conditions fixed in the parametric analysis

Case	Value of the inlet parameters				
	T_{al} °C	m_a kg/s	y_{al} kg_{wv}/kg_a	T_{s1} °C	X_1 kg_d/kg_s
Desiccant Inlet Concentration (X_1)	33	0.037	0.0221	32	0.302-0.400
Desiccant Inlet Temperature (T_{s1})	33	0.037	0.0221	28-38	0.302
Inlet Air Humidity Ratio (y_{a1})	33	0.037	0.0175-0.020	32	0.302
Inlet Air Mass Flow Rate (m_a)	33	0.037-0.047	0.0221	32	0.302
Inlet Air Temperature (T_{a1})	28-34	0.037	0.0221	32	0.302

3.4. Effect Of Inlet Parameters To Performance Parameters

After completing all parametric analysis on the previous section, one can observe that inlet humidity ratio is the most influencing parameter for moisture removal rate, MRR (52.68%) and dehumidification effectiveness (48.59%). The second goes to inlet air temperature, which caused 30.17% increment in average, while 22.68% increment on the dehumidification effectiveness. Other inlet parameters such as the concentration of inlet liquid desiccant, desiccant inlet temperature and inlet air mass flow rate did increase or decrease the performance parameters but only in a less significant ways, in average less than 5%. However, both inlet humidity ratio and inlet air temperature depending on the outdoor climate condition and it is uncontrollable. Therefore, this combined solar chimney and liquid desiccant plant can perform well in place with high humidity ratio and high ambient temperature.

3.5. Improvement Strategies

A summary of the strategies suggested has been tabulated as shown in table 3.2 below:

Table 3.2. Improvement strategies to enhance the performance of the dehumidifier.

Opportunities	How
The concentration of Liquid Desiccant	Increase the Concentration from 30.2% to 40%. As proven in the parametric in the previous section, increase the concentration may results in increment in the moisture removal rate, MRR.
Type of Liquid Desiccant	In this study, Calcium Chloride is used as the liquid desiccant solution. Lithium Bromide (LiBr), Lithium Chloride (LiCl) and even mixture of $CaCl_2$ and LiCl can be used as liquid desiccant for combined solar chimney and liquid desiccant plant.[14]
Size of the dehumidifier	Increase the size of the dehumidifier by increasing it thickness[7]. However, the equilibrium point between pressure loss and moisture removal rate should be found to obtain the best results.

4. Conclusion

The conclusions of this study are as below:

- A combined solar chimney and liquid desiccant plant micro-plant has potential to be developed into a new technology for simultaneous production of fresh water and electricity in the near future.
- The highest dehumidification achieved is around 22.78% (24Oct) as shown in figure 3.1, while the highest moisture removal rate (MRR) registered among the five testing days are 0.22ml/s (28Oct) as shown in figure 3.2. It can be conclude that the highest MRR does not result in the highest effectiveness of the dehumidifier. Dehumidification effectiveness and MRR does not depend on the same parameters but a combination of a few parameters, such as ambient temperature, relative humidity and even the climate condition.
- After completing the parametric analysis (refer to table 3.1), it is found that the most influencing parameter is the inlet air humidity ratio although it is uncontrollable and fully depends on the outdoor conditions.
- Several improvement strategies had been suggested for future studies. (refer to table 3.2)

5. Acknowledgment

The authors would like to express their sincere gratitude to the Ministry of Higher Education, Malaysia for financing this work through the grant FRGS 1632. The Research Management Centre, Universiti Tun Hussein Onn Malaysia is also acknowledged for their continuous support and cooperation.

IOP Conf. Series: Materials Science and Engineering **824** (2020) 012009 doi:10.1088/1757-899X/824/1/012009

6. References

[1] E. Delyannis 2003 Historic background of desalination and renewable energies *Sol. Energy*. vol. **75** no. 5 pp. 357–366

[2] Solar Chimney [Accessed: 20-May-2018] Available: https://www.sbp.de/en/project/solar-chimney-pilot-plant-manzanares/

[3] W. HAAF 1984 Solar Chimneys *Int. J. Sol. Energy*. vol. **2** no. 2 pp. 141–161

[4] W. Haaf, K. Friedrich, G. MAYR†, and J. Schlaich 1983 Part I: Principle and Construction of the Pilot Plant in Manzanares *Int. J. Sol. Energy*. vol. **2** no. 1 pp. 3–20

[5] Sabah K.Mahal and Ahmad J.Alimin 2016 A review of the hybrid solar chimney and water desalination systems *I.RE.M.E.* vol. **10** no 6 pp. 419–436

[6] M. M. Bassuoni 2014 A simple analytical method to estimate all exit parameters of a cross-flow air dehumidifier using liquid desiccant *J. Adv. Res.* vol. **5** no. 2 pp. 175–182

[7] W. Z. Gao, J. H. Liu, Y. P. Cheng, and X. L. Zhang 2012 Experimental investigation on the heat and mass transfer between air and liquid desiccant in a cross-flow dehumidifier *Renew. Energy*. vol. **37** no. 1 pp. 117–123

[8] S. M. Pineda and G. Diaz 2011 Performance of an adiabatic cross-flow liquid-desiccant absorber inside a refrigerated warehouse *Int. J. Refrig.* vol. **34** no. 1 pp. 138–147.

[9] D. Babakhani, M. Soleymani, and A. Moheb 2010 Heat and mass transfer between air and liquid desiccant in cross-flow contact systems *Chem. Eng. Technol.* vol. **33** no. 2 pp. 281–291

[10] C. G. Moon, P. K. Bansal, and S. Jain 2009 New mass transfer performance data of a cross-flow liquid desiccant dehumidification system *Int. J. Refrig.* vol. **32** no. 3 pp. 524–533

[11] X. H. Liu, Y. Zhang, K. Y. Qu, and Y. Jiang 2006 Experimental study on mass transfer performances of cross flow dehumidifier using liquid desiccant vol. **47** pp. 2682–2692.

[12] Sabah K.Mahal and Ahmad J.Alimin 2017 Advancement in the technology of liquid desiccant systems. *I.RE.M.E.* vol. **11** no 3 pp. 191–199

[13] H. E. Gad, A. M. Hamed, and I. I. El-Sharkawy 2001 Application of a solar desiccant/collector system for water recovery from atmospheric air *Renew. energy*. vol. **22** no. 4 pp. 541–556.

[14] A. Pesaran 1993 A review of desiccant dehumidification technology *ERPI's Electr Dehumidification Energy Effic. Humidity Control Commer. Industitutional Build. Conf*

IOP Conf. Series: Materials Science and Engineering **824** (2020) 012010 doi:10.1088/1757-899X/824/1/012010

Preliminary SWCNTs-kerosene on Carreau MHD flow over a wall with distinct thickness

Jaafar Abdul Abbas Abbood Al-Nasrawi*, Rozaini Bin Roslan, Owoyemi Abiodun Ezekiel, Radiah Mohamad

Faculty Faculty of Applied Sciences and Technology,
University Tun Hussein Onn Malaysia, 86400 Parit Raja, Batu Pahat, Johor, Malaysia.

E-mail: * jaafarali2011@gmail.com

Abstract. The Magnetohydrodynamics (MHD) Carreau over a stretched surface on a kerosene oil based is examined in this paper. Most specifically, we extended and reported the effects of various parameter namely; magnetic field, thermal radiative and heat generation on the temperature. Thermal conductivity enhancements of water and seawater in the presence of single-walled carbon nanotubes (SWCNTs) are presented. We used transformation of similarity to reduce and regenerated the partial differential equations (PDEs) into ordinary differential equations (ODEs) for the nonlinear governing equation alongside with their related boundary conditions. The result of the regenerated ODEs is solved numerically by the application of the fourth/fifth order Runge Kutta Fehlberg technique with method of shooting using Maple code. The result obtained indicated that there is significant positive correlation between high electric field. The result reported that the temperature parameter is decreased with all parameter when the electric is high (El=0.5), which indicated that the electric filed is a dominant of SWCNTs-kerosene.his study was conducted to develop and study the performance of a Photovoltaic Thermal (PVT) system technology, a system which combines the Photovoltaic (PV) system with a solar thermal (T) system that produces electricity and hot water simultaneously. The objective of this study is to develop a working prototype of the PVT system for roofing system application with water as the cooling agent. A polycrystalline PV cells was used as the solar collector while a transparent rubber tube was attached at the backside of the PV panel with water flowing directly from a water tap source. Two designs were examined for the heat absorber design which was categorized as the parallel and spiral flow design. The system was run at an open space to simulate real life application. A working PVT system was developed; however an electrical efficiency drops 0.3% and 1.07% on the parallel and spiral design respectively when compared to the standard PV system.

1. Introduction

Stephen Choi, in 1995, introduced the term nanofluid [1]. The term is described as a liquid that contain dispersion of submicronic solid particles and the particle that the contains the nano is called nanoparticles. The fundamental objective of using of nanoparticle is to diffuse solid particles in the fluid to boost thermal conductivity. In real-time application, nanofluid has been reported to be boosting the transfer of heat than 50 percent especially when the nanoparticle volume ratio of the base-oil is below 0.3 percent [2]. The apparent (surface) region to the ration of the volume of nanoparticle is very huge. This however, fetches a huge and enormous propelling pressure for diffusion, most specifically when the temperature is raised. When comparing nanofluid with other based fluid, nanofluid has an upper hand about boosting in thermal conductivity [3], [4], [5], [6].

Heat transfer enhancement through modification of fluid thermo-physical properties by adding dispersed particles to a base fluid has been a topic of interest since the early works by Ahuja as indicated in [7]. With the new discovery in thermal conductivity, many researchers have carried out research activities to improve the thermal conductivity properties so that it would behave same as fluid but has thermal conductivity as metal by adding conductive solids into fluids, [4], [8], [5], [6], [3]. Nanoparticle used are usual made of metal (Al, Cu), [9], oxides (Al_2O_3) [10] and SWCNTs [11] just to mention a few. Therefore, because of the new additive it is very import and necessary to explore more on the impact of heat transfer behaviour and characteristics, namely, thermal conductivity, heat transfer, viscosity and so on.

In this research direction, mathematicians, physicists and engineers are among the academic researchers who contribute to the knowledge of thermal conductivity [12], [13], [1]. Though, research has shown that nanofluid provides higher transfer of heat with respect of the based fluid [14], [15]. [16] but regarding the kerosene-based research direction, it is observed that only scanty research have been reported with respect to the nanofluid thermal particles [17].Various nanoparticle shapes have been utilized over the experiment [18], [19], which presented the first kind of experimental analysis. The thermal conductivity improvement is due to the shape of the included nanoparticles into suspension. [20] analyzed the role of the silicon carbide (SiC) nanoparticles where they admitting the shape affirm disc or platelet and consistently diffused in water on mechanical assets and thermal conductivity enrichment.

Recent, report reviewed that one of the essential outputs of the existences of the powerful electrical conductivity promotes a distinguishable and distinctive subject matter on the fluid SWCNTs flow. In industrial applications, nanoparticles have been used to avoid overheat of the instrument [21]. Thus, nanoparticle play an important role as the coolant substance to transfer heat example of such is our computer system [22]. The heat sink helps to absorb and cool the heat generated in the system, thereby avoid being overheated. This is because the diameter less than 100nm and this condition will affects the properties of the liquid surface area to enhance the heat transfer compare to the micron-size particle [23]. As far as development in technology and industries remain, the exchange of thermal conductivity must be updated time to time. The old method thermal conductivity, which is referred to as traditional means of transferring heat fluid such as Maxwell theories [24] has allot of difficulties in improving the cooling capacities [25], [26], [27], and is not sufficient or cannot explain behaviour of the thermal conductivity of nanoparticle. However, recently, effort has been carried out to improve and report the thermal conductivity's property by making an adding into the based fluid [28]. A few shapes of the nanoparticles that have been analyzed include cylindrical-shape [9], [18], and shuttle-like shape [29].

In this paper, our aim is to extend and explore more on the effects of various parameter namely; magnetic field, thermal radiative and heat generation on the temperature and to report the effects of various parameters in thermal conductivity of the nanofluid to kerosene based SWCNTs.
Methodology

2. Mathematical formation

The aim of this work is to clarify several aspects of a mathematical form and report Carreau SWCNTs – kerosene flow of nanofluid attributed of nonlinear of the stretching with extensible depth. $U_w = U_o.(x+b)^m$ is the velocity used in the paper to construct the stretching sheet, where U_0 represents the source velocity for the constructed sheet, $B.(x+b)^{(1-m)/2}$ is set to be the thickness of sheet, b, m are set to be its constant dimensionless and the rate of power law directory, respectively. The review of the model is constructed to be pleased only for $m \neq 1$, the reason is that of for $m = 1$, the issue lessen the flat-sheet. Flexible strength of the magnetic field $B_0.(x)$ is vertically imposed to the plate. On the other hands, the electrical and field of the magnetic is perpendicular to the surface which deferent thickness. Fig. 1 displayed the physical configuration of SWCNTs – kerosene nanofluid as used in this paper.

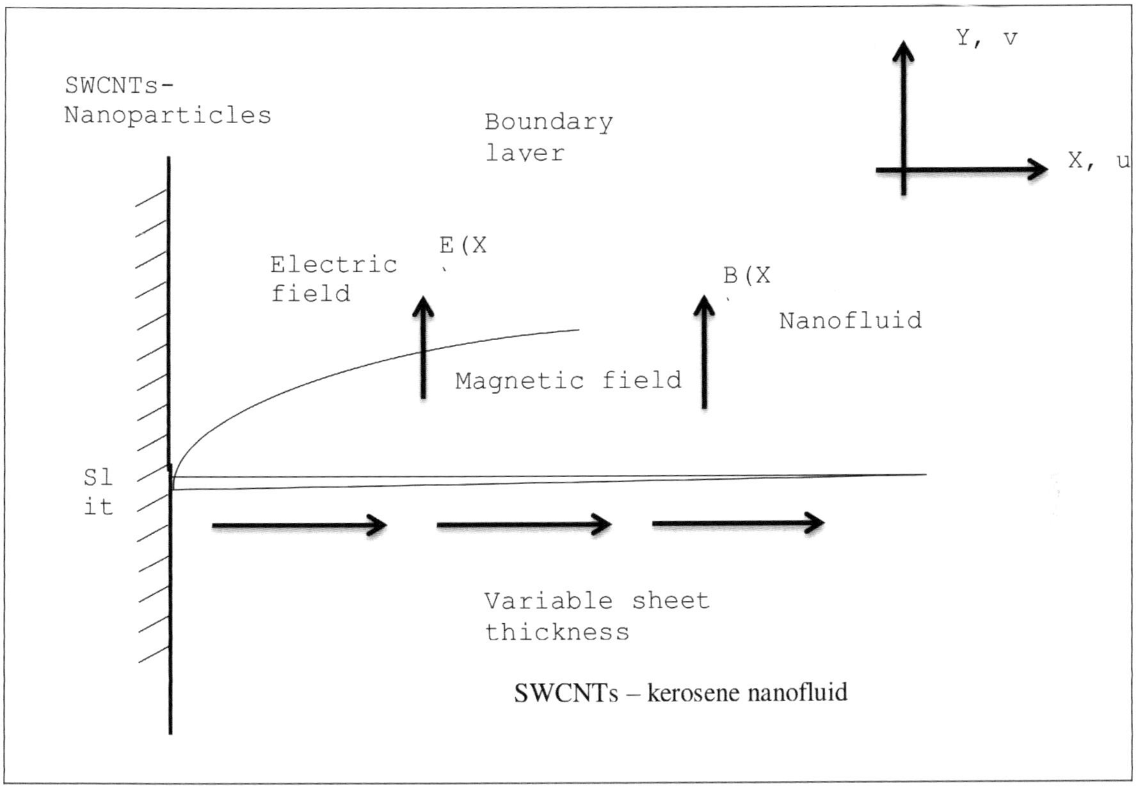

Figure 1. Physical configuration of SWCNTs – kerosene nanofluid system.

System based on the boundary layer approximation, equations (1) - (4) can be derived:

$$\frac{\partial u}{\partial x}+\frac{\partial v}{\partial y}=0 \tag{1}$$

$$u\frac{\partial u}{\partial x}+v\frac{\partial u}{\partial y}=\frac{\mu_{nf}}{\rho_{nf}}\left(\frac{\partial^2 u}{\partial y^2}+\frac{3(n-1)}{2}\Gamma^2\left(\frac{\partial u}{\partial y}\right)^2\frac{\partial^2 u}{\partial y^2}\right)-\frac{\sigma_{nf}}{\rho_{nf}}\left(E.(x)B(x)-B_0^2(x).u\right) \tag{2}$$

$$u\frac{\partial T}{\partial x}+v\frac{\partial T}{\partial y}=\frac{k_{nf}}{(\rho c_p)_{nf}}\left(\frac{\partial^2 T}{\partial y^2}\right)-\frac{1}{(\rho c_p)_{nf}}\frac{\partial q_r}{\partial y}+\frac{1}{(\rho c_p)_{nf}}\tau\left[(D_B)_{nf}\left\{\frac{\partial C}{\partial y}\frac{\partial T}{\partial y}\right\}+\frac{(D_T)_{nf}}{T_\infty}\left\{\left(\frac{\partial T}{\partial y}\right)^2\right\}\right]+\frac{Q_0(T-T_\infty)}{(\rho c_p)_{nf}} \tag{3}$$

$$\frac{\partial C}{\partial x}+v\frac{\partial C}{\partial y}=(D_B)_{nf}\left(\frac{\partial^2\phi}{\partial y^2}\right)+\frac{(D_T)_{nf}}{T_\infty}\left\{\frac{\partial^2 T}{\partial y^2}\right\}-K(C-C_\infty) \tag{4}$$

Subject to the boundary conditions

$$u\left(x, B(x+b)^{(1-m)/2}\right) = U_o(x+b)^m;\ v\left(x+B(x+b)^{(1-m)/2}\right) = 0;\ T(x+B(x+b)^{(1-m)/2}) = T_w\, C[x+B(x+b)^{(1-m)/2}] = C_w;\ u(x,\infty) \to 0;\ T \to T_\infty;\ C \to C_\infty\, as\ y \to \infty \quad (5)$$

TABLE 1. Nomenclature.

Parameters	Symbols	parameters	Symbols
Magnetic field factor	$\boldsymbol{B(X) = B_O(x+b)^{\frac{1-m}{2}}}$	electrical field factor	$\boldsymbol{E(X) = E_O(x+b)^{\frac{1-m}{2}}}$
Electrical conductivity	σ	nanofluid temperature	T
x and y Velocities	u,v	Nanofluidconcentration	C
Concentration of the nanofluidnear the wall	C_w	Ambient concentration of nanofluid	C_∞
Nanofluid effiective density	ρ_{nf}	Nanofluid thermal diffusivity	α_{nf}
Nanofluid dynamic viscosity	μ_{nf}	nanofluid heat capacitance	$(\rho c_p)_{nf}$

Table 1 displayed the parameters with its respective symbols used in this article for ease identification

$$\rho_{nf} = (1-\zeta)\rho_f + \zeta\rho_s,\ \mu_{nf} = \frac{\mu_f}{(1-\zeta)^{2.5}},\ (\rho\beta)_{nf} = (1-\zeta)(\rho\beta)_f + \zeta(\rho\beta)_s$$

$$(\rho c_p)_{nf} = (1-\zeta)(\rho c_p)_f + \zeta(\rho c_p)_s\quad \alpha_{nf} = \frac{k_{nf}}{(\rho c_p)_{nf}},\ \frac{k_{nf}}{k_f} = \left\{\frac{(k_s+(l-1)k_f) - (l-1)\zeta(k_f-k_s)}{(k_s+(l-1)k_f) + \zeta(k_f-k_s)}\right\},$$

$$\sigma_{nf} = (1-\zeta)\sigma_f + \zeta\sigma_s\ (D_B)_{nf} = (1-\zeta)(D_B)_f \quad (6)$$

The Maxwell design method was used to determine the dynamic thermal or electrical conductivities of liquid-solid suspensions. Let σ_f and σ_s-be the base fluid and nanoparticle electrical conductivities, k_f and k_s are the base-fluid, and nanoparticle thermal conductivities, ζ is the volume fraction of nanoparticle, ρ_f and ρ_s are the base fluid and nanoparticle densities, σ_f and σ_s are the based fluid, and nanoparticle electrical conductivities, μ_f is the base-fluid dynamic viscosity, k_{nf} is the nanofluid effective thermal conductivity, $q_r = -\frac{4\sigma^* \partial T^4}{3k^* \partial y}$ [30] is the Rosseland's dissipation for the radiative heat flux, k^* is the Rosseland signify absorption coefficient and σ^* is the Stephen-Boltzmann constant. Here the term T^4 is linearized using Taylor's series about T_∞ as $T^4 = 4T_\infty^3 T - 3T_\infty^4$. The similarity conversions were performed as below:

$$\psi=\left[\frac{2\nu_f U_0(x+b)^{m-1}}{m+1}\right]^{\frac{1}{2}} f(\eta),\eta=\left[\frac{(m+1)U_0(x+b)^{(m-1)}}{2\nu_f}\right]^{\frac{1}{2}} y,\theta(\eta)=\frac{T-T_\infty}{T_w-T_\infty},\phi(\eta)=\frac{C-C_\infty}{C_w-C_\infty}, \tag{7}$$

where ν_f is the fluid kinematic viscosity and ψ is the function $u=\frac{\partial\psi}{\partial y}, v=\frac{\partial\psi}{\partial x}$.

Therefore, the dimensionless form of Equations (1) - (4) to as follow

$$F'''\left(1+\left(\frac{3(n-1)}{2}W^2(F'')^2\right)\right)+G.B\left(\frac{2m}{m+1}F'^2-FF''\right)+G.A.M(E_1-F'^2)=0 \tag{8}$$

$$\theta''+\Pr_f.E.\left(F\theta'+(A1)Nb\theta'\phi'+(A1)Nt\theta'^2+\lambda\theta\right)=0 \tag{9}$$

$$\phi''+(A1).\left(\frac{1}{(A1)}\frac{Nt}{Nb}\theta''+LeF\theta'-Le\gamma\phi\right)=0 \tag{10}$$

Subject to the boundary conditions:

$$F(\alpha)=\alpha\frac{1-m}{1+m}, F'(\alpha)=1, \theta(\alpha)=1, \phi(\alpha)=1; F'(\infty)=0, \theta(\infty)=0, \phi(\infty)=0 \tag{11}$$

$$K_{nf}=(k_s+(e-1)(k_f)-(e-1)\zeta(k_f-(k_s), K_f=(k_s+(e-1)(k_f)-\zeta(k_f-(k_s), B=\left(1-\zeta+\zeta\frac{\rho_s}{\rho_f}\right), T=\frac{K_{nf}}{K_f}, \tag{12}$$

$$E=\left(1-\zeta+\zeta\frac{\rho_s(cp)_s}{\rho_f(cp)_f}\right), A=\left(1+\frac{3\left(\frac{\sigma_s}{\sigma_f}-1\right)\zeta}{\left(\frac{\sigma_s}{\sigma_f}+2\right)-\left(\frac{\sigma_s}{\sigma_f}-1\right)\zeta}\right), G=(1-\zeta)^{2.5}, \tag{13}$$

Here, $\alpha=B\sqrt{\frac{U_0(m+1)}{2\nu_f}}$ -is the depth parameter of the corresponding wall and $\eta=B\sqrt{\frac{U_0(m+1)}{2\nu_f}}$ -is the surface dependent plate. By substituting, $f(\alpha)=f(\eta-\alpha)=f(\eta)$, $\theta(\alpha)=\theta(\eta-\alpha)=\theta(\eta)$, and $\phi(\alpha)=\phi(\eta-\alpha)=\phi(\eta)$, the above equations can be rewritten as:

$$f'''\left(1+\left(\frac{3(n-1)}{2}W^2(f'')^2\right)\right)+G.B\left(\frac{2m}{m+1}f'^2-ff''\right)+G.A.M(E_1-f'^2)=0 \tag{14}$$

$$\theta'' + \Pr_f.E.\left(f\theta' + (A1)Nb\theta'\phi' + (A1)Nt\theta'^2 + \lambda\theta\right) = 0 \tag{15}$$

$$\phi'' + (A1).\left(\frac{1}{(A1)}\frac{Nt}{Nb}\theta'' + Lef\theta' - Le\gamma\phi\right) = 0 \tag{16}$$

with boundary conditions:

$$f(\alpha) = \alpha\frac{1-m}{1+m}, f'(\alpha) = 1, \theta(\alpha) = 1, \phi(\alpha) = 1; f'(\infty) = 0, \theta(\infty) = 0, \phi(\infty) = 0 \tag{17}$$

where, $M = \frac{2\sigma_f B_0^2}{U_0(m+1)(x+b)^{m-1}\rho_f}$ - the magnetic parameter, $W = (m+1)\frac{U_0^3(x+b)^{3m-1}}{2\nu_f}\Gamma^2$ -is the Weissenberg number, $\Pr_f = \frac{\Pr}{\left(\frac{k_{nf}}{k_f} + \frac{4}{3}R\right)}$ - is the effective Prandtl number, $\Pr = \frac{(\mu c_p)_f}{k_f}$ is the Prandtl-number, $\gamma = \frac{2K}{U_0(m+1)(x+b)^{m-1}}$ -is the chemical reaction, $\mathrm{Re}_x = \frac{xU_0(x)}{\nu_f}$ - is the local Reynolds number, $E_1 = \frac{E_0}{B_0U_0(x+b)^m}$ - is the electric parameter, $N_b = \frac{(\rho c)_s(D_B)_f(C_w - C_\infty)}{(\rho c)_f\nu_f}$ - is the Brownian motion parameter, $N_t = \frac{(\rho c)_s(D_T)_f(T_w - T_\infty)}{(\rho c)_f\nu_f T_\infty}$ - is the thermophoresis parameter, $Le = \frac{\nu_f}{(D_B)_f}$ - is the Lewis number and $R = -\frac{4\sigma^* T_\infty^3}{k_f k^*}$ -is the solar radiation-conduction parameter.

$$\mathrm{Re}_x^{\frac{1}{2}} C_f = \frac{1}{(1-\zeta)^{2.5}}\left(\frac{m+1}{2}\right)^{\frac{1}{2}} f''(0),\ \mathrm{Re}_x^{-\frac{1}{2}} Nu_x = -\left(\frac{k_{nf}}{k_f} + \frac{4}{3}R\right)\left(\frac{m+1}{2}\right)^{\frac{1}{2}}\theta'(0)\ \ Sh_x \mathrm{Re}_x^{-\frac{1}{2}} = -\frac{1}{(1-\zeta)}\left(\frac{m+1}{2}\right)^{\frac{1}{2}}\phi'(0) \tag{18}$$

3. Result and discussion

In this section, the fourth/fifth order Runge Kutta method is utilised along with shooting method for the approximate solutions of equation (14) – (16) with the boundary condition in equation (17). The method is introduced by Runge Kutta around 1900 and has seen been used by different authors [9], [16]. The simulation is computed for the various parameters effects on temperature, $-\boldsymbol{\theta}'(\mathbf{0})$, along with electric field.

Fig. 2 shown the nanoparticle shapes of the nanoparticles. Table 2 displayed the thermo-physical resources of the fluid and nanoparticles that were used in this work. Table 3-5 displayed the several parameters effects in the temperature profiles obtained for the dimensionless form for El = 0.0 and El = 0.5.

The effects of parameters on the magnetic field with M = 0.10, M= 0.50 and M = 0.90 is presented in Fig. 3. It is interested to report that the temperature, $(-\theta'(0)\ )$ decreased with the extension and increased of magnetic field. This is because of the random motion of particles suspended in a fluid. Table 3 presents that the increase in the magnetic field is resulted with an increase in the temperature, $(-\theta'(0)\ )$. As it can be seen, as the magnetic field is increasing then the concentration on the temperature, $(-\theta'(0)\ )$ is increase when El = 0.0 and as when El = 0.5. This means temperature gradient is not influenced compared to the temperature fluctuation in the direction vertical to the magnetic field.

IOP Conf. Series: Materials Science and Engineering **824** (2020) 012010 doi:10.1088/1757-899X/824/1/012010

Fig. 4 presented the effects of parameters on the thermal radiation with R = 0.1, R = 1 and R = 3. It is observed that the temperature circulation for all the nanofluids increased and the rate of transfer in the heat concentration increased with the extension and increase of thermal radiation strength. The boundary of the thermal surface depth of SWCNTs – kerosene oil is dominant. This is due to the combined effects of thermal conductivity and density of the SWCNTs nanofluid. As it can be seen, as the thermal radiation strength is increasing then the concentration on the temperature, $(-\theta'(0)\ )$ is decreasing when El = 0.0 and as when El = 0.5. Table 4 presents that the temperature circulation for all the nanofluids increased but in contrast the rate of transfer in the heat concentration increase of thermal radiation strength. Particularly, between boundary layer at when $R = 1.0, R = 3.0$ and $R = 5.0$.

It is indicated from Fig. 5, which is the parameters' effects on the heat generation (γ) with γ = 0.1, γ = 0.3 and γ = 0.5. It is observed that the temperature circulation for all the nanofluids increase but in contrast the rate of transfer in the heat concentration increased with increase of the heat generation (γ). It is well known that MHD pressure drop reduction is an important issue in the fluid mechanics. The MHD pressure drop is caused by the heat generation of a resistive electromagnetic force, which in turn increases the friction force because of the resultant flattened velocity distribution due to induced by Lorentz force as discussed in [31]. Table 5 presents that temperature circulation for all the nanofluids increase but in contrast the rate of transfer in the heat concentration increase of the heat generation (γ). Particularly, the boundary of the thermal surface depth of SWCNTs – kerosene oil is dominant. As it can be seen, as the heat generation (γ) strength is increasing then the concentration on the temperature, $(-\theta'(0)$ is decreasing when El = 0.0 and as when El = 0.5.

The temperature, $(-\theta'(0)\ )$ of the nanofluids decrease in the presence of the magnetic field (M = 0.1, M= 0.5 and M = 0.9), with the increase in other nanofluid parameters, the thermal radiation (R) and heat generation (γ). Particularly, thermal boundary layer thickness of SWCNTs – kerosene is stronger.

Table 2. Thermo-physical resources of the fluid and nanoparticles.

	$\sigma(\Omega^{-1}m^{-1})$	$\rho(kg/m^3)$	$c_p(J/kgK)$	$k(W/mK)$
Kerosene	783	2090	0.145	4.86
SWCNTs-kerosene	2600	42.5	6600	1.26

Sphere (m =3.0) Cylinder (m=6.3628) Lamina (m=16.1576)

Figure 2. Nanoparticle shapes.

IOP Conf. Series: Materials Science and Engineering **824** (2020) 012010 doi:10.1088/1757-899X/824/1/012010

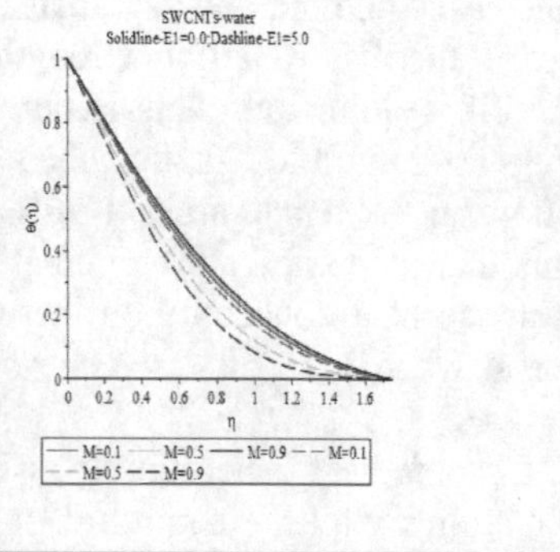

Figure 3. Magnetic field (M)

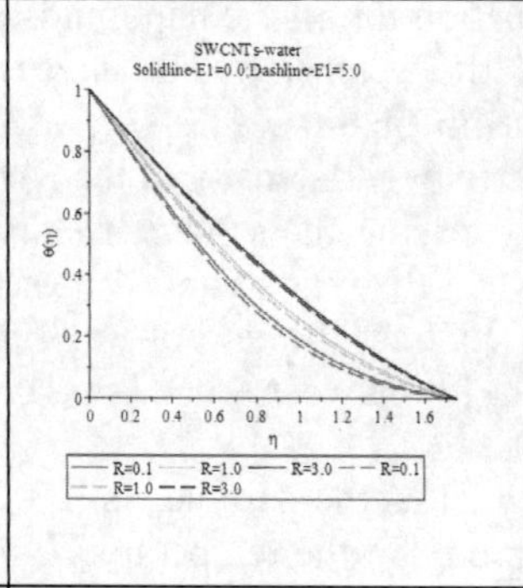

Figure 4. Thermal Radiation(R).

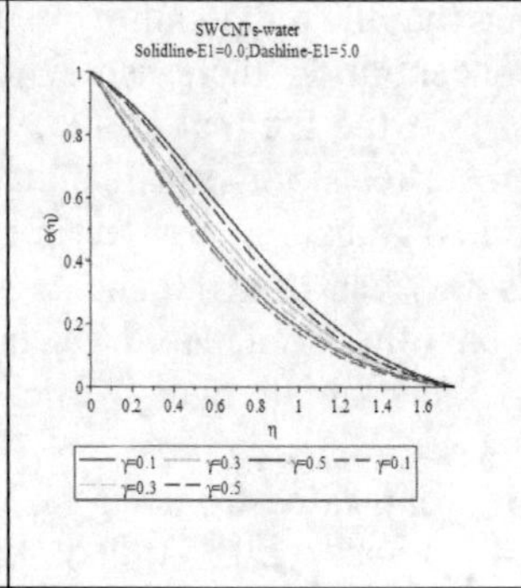

Figure 5. Heat generation (γ)

Table 3. Magnetic field deposition on $f''(0)$, $-\theta'(0)$and $-\varphi'(0)$.

M	$f''(0)$	$-\theta'(0)$	$-\varphi'(0)$	Remark
0. 1	-0.061924360.974637508	0.135958934		E1 = 0.0
0.5	-0.217986560.952379683	0.151368698		
0.9	-0.354230260.933137979	0.164942024		
0. 1	0.206321950	1.014838665	0.109158238	
0.5	1.121600764	1.139776354	0.034226946	
E1 = 5.0				
0.9	2.014618856	1.244032672	0.016290443	

Table 4. Thermal Radiation deposition on $f''(0)$, $-\theta'(0)$ and $-\varphi'(0)$.

R	$f''(0)$ Remark	$-\theta'(0)$	$-\varphi'(0)$	
0. 1	-0.061924360.974637508	0.135958934		E1 = 0.0
1.0	-0.061924370.833665065	0.349542219		
3.0	0.061924415	0.716012830	0.533825944	
0. 1	0.206321950	1.014838665	0.109158238	
1.0	0.206321932	0.863339187	0.331765380	E1
= 5.0				
3.0	0.206321895	0.733844123	0.526822455	

Table 5. Heat generation deposition on $f''(0)$, $-\theta'(0)$ and $-\varphi'(0)$.

γ	$f''(0)$	$-\theta'(0)$	$-\varphi'(0)$	Remark
0. 1	-0.061924366	0.974637508	0.1359589349	El = 0.0
0.3	-0.080758717	0.710446099	0.1653936047	
0.5	-0.061924363	0.396259884	0.1956959508	
0. 1	0.2063219507	1.014838665	0.1091582385	
0.3	0.2063219526	0.768525605	0.1304137086	El = 5.0
0.5	0.2063219546	0.470933108	0.1562449437	

To confirm the present article with previous article such as [32], we compare the chemical reaction in [32], which is validate for this parameter. The result with the theoretical solution for the velocity profile is observed to be very good. Fig. 6 displayed the comparison profile.

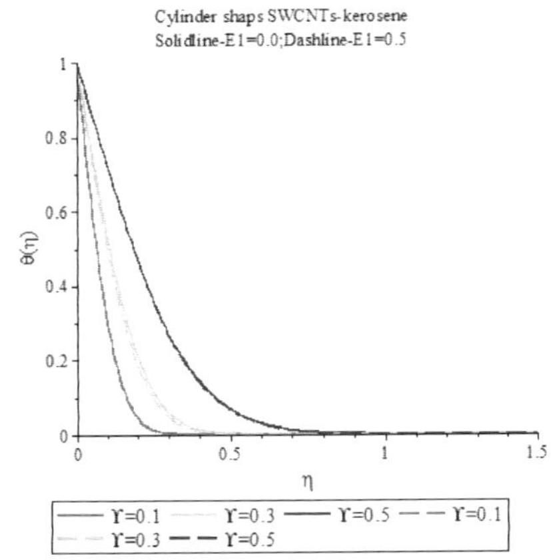

Figure 6. Comparison of heat generation effects on temperature with [32].

4. Conclusion

The effects of various parameter namely; the thermal radiation (R) and heat generation (γ), on temperature are investigated in this work. Graphically, the results are presented and the conclusion is drawn on the various effect on the temperature with electric field.

The boundary conditions of the ODEs were solved using the same value of the governing parameter. We took several variable of the parameter as manipulated and other parameters as the constant variable, we were able to get a lot of results and reactions, which include;

- In the capacity of magnetic on the profile, the velocity decreased when the parameter M is increase, whereas at the thickness of the boundary layer for El = 0.5 is weaker as compared to El = 0.0. There is no significance difference in temperature and concentration with increasing M.
- From the profile boundary layer thickness, is indicated that the thermal radiation, heat generation and chemical is large which is show that the parameter has high effect on temperature.

- The parameter of the temperature is not increase with all the parameter when the electric field is at (El=0.5). This show that the electric field is dominate of kerosene-SWCNT.

5. References

[1] S. U. S.Choi and J. A. Eastman, "Thermal Conductivity of Fluids With.Pdf," *American Society of Mechanical Engineers*. pp. 99–105, 1995.
[2] Y. J. Hwang *et al.*, "Investigation on characteristics of thermal conductivity enhancement of nanofluids," *Curr. Appl. Phys.*, vol. 6, no. 6 SPEC. ISS., pp. 1068–1071, 2006.
[3] R. Kandasamy, R. Dharmalingam, and K. K. S. Prabhu, "Thermal and solutal stratification on MHD nanofluid flow over a porous vertical plate," *Alexandria Eng. J.*, vol. 57, no. 1, pp. 121–130, 2018.
[4] A. E. Kabeel, E. M. S. El-Said, and S. A. Dafea, "A review of magnetic field effects on flow and heat transfer in liquids: Present status and future potential for studies and applications," *Renew. Sustain. Energy Rev.*, vol. 45, pp. 830–837, 2015.
[5] P. Singh and M. Kumar, "Mass transfer in MHD flow of alumina water nanofluid over a flat plate under slip conditions," *Alexandria Eng. J.*, vol. 54, no. 3, pp. 383–387, 2015.
[6] P. Sudarsana Reddy, A. J. Chamkha, and A. Al-Mudhaf, "MHD heat and mass transfer flow of a nanofluid over an inclined vertical porous plate with radiation and heat generation/absorption," *Adv. Powder Technol.*, vol. 28, no. 3, pp. 1008–1017, 2017.
[7] L. G. Asirvatham, "Nanofluid heat transfer and applications," vol. 1, no. 2, pp. 113–115, 2015.
[8] N. Sandeep, C. Sulochana, and B. Rushi Kumar, "Unsteady MHD radiative flow and heat transfer of a dusty nanofluid over an exponentially stretching surface," *Eng. Sci. Technol. an Int. J.*, vol. 19, no. 1, pp. 227–240, 2016.
[9] R. Kandasamy, N. A. bt Adnan, and R. Mohammad, "Nanoparticle shape effects on squeezed MHD flow of water based Cu, Al2O3 and SWCNTs over a porous sensor surface," *Alexandria Eng. J.*, 2016.
[10] N. Freidoonimehr, M. M. Rashidi, and S. Mahmud, "Unsteady MHD free convective flow past a permeable stretching vertical surface in a nano-fluid," *Int. J. Therm. Sci.*, vol. 87, pp. 136–145, 2015.
[11] R. Dharmalingam, R. Kandasamy, and K. K. Sivagnana Prabhu, "Lorentz forces and nanoparticle shape on water based Cu, Al2O3 and SWCNTs," *J. Mol. Liq.*, vol. 231, pp. 663–672, 2017.
[12] H. Xie, J. Wang, T. Xi, Y. Liu, F. Ai, and Q. Wu, "Thermal conductivity enhancement of suspensions containing nanosized alumina particles," *J. Appl. Phys.*, vol. 91, no. 7, pp. 4568–4572, 2002.
[13] R. Prasher, P. E. Phelan, and P. Bhattacharya, "Effect of aggregation kinetics on the thermal conductivity of nanoscale colloidal solutions (nanofluid)," *Nano Lett.*, vol. 6, no. 7, pp. 1529–1534, 2006.
[14] M. Hemmat, W. Yan, M. Akbari, A. Karimipour, and M. Hassani, "Experimental study on thermal conductivity of DWCNT-ZnO / water-EG nano fl uids ☆," *Int. Commun. Heat Mass Transf.*, vol. 68, pp. 248–251, 2015.
[15] J. Raza, A. M. Rohni, and Z. Omar, "MHD flow and heat transfer of Cu–water nanofluid in a semi porous channel with stretching walls," *International Journal of Heat and Mass Transfer*, vol. 103. pp. 336–340, 2016.
[16] R. Kandasamy, R. Mohammad, N. A. B. M. Zailani, and N. F. B. Jaafar, "Nanoparticle shapes on squeezed MHD nanofluid flow over a porous sensor surface," *J. Mol. Liq.*, vol. 233, pp. 156–165, 2017.
[17] S. E. B. Maïga, C. T. Nguyen, N. Galanis, G. Roy, T. Maré, and M. Coqueux, "Heat transfer

enhancement in turbulent tube flow using Al2O3 nanoparticle suspension," *Int. J. Numer. Methods Heat Fluid Flow*, vol. 16, no. 3, pp. 275–292, 2006.

[18] H. Xie, J. Wang, T. Xi, and Y. Liu, "Thermal Conductivity of Suspensions Containing Nanosized SiC Particles," *Int. J. Thermophys.*, vol. 23, no. 2, pp. 571–580, 2002.

[19] S. M. S. Murshed, K. C. Leong, and C. Yang, "Enhanced thermal conductivity of TiO2- Water based nanofluids," *Int. J. Therm. Sci.*, vol. 44, no. 4, pp. 367–373, 2005.

[20] D. Singh *et al.*, "An investigation of silicon carbide-water nanofluid for heat transfer applications," *J. Appl. Phys.*, vol. 105, no. 6, 2009.

[21] K. V. Wong and O. De Leon, "Applications of nanofluids: Current and future," *Advances in Mechanical Engineering*, vol. 2010. 2010.

[22] M. R. Islam, B. Shabani, and G. Rosengarten, "Electrical and Thermal Conductivities of 50/50 Water-ethylene Glycol Based TiO2 Nanofluids to be Used as Coolants in PEM Fuel Cells," in *Energy Procedia*, 2017, vol. 110, pp. 101–108.

[23] W. Anderson, D. Kozak, V. A. Coleman, Å. K. Jämting, and M. Trau, "A comparative study of submicron particle sizing platforms: Accuracy, precision and resolution analysis of polydisperse particle size distributions," *J. Colloid Interface Sci.*, 2013.

[24] M. Sheikholeslami, D. D. Ganji, and M. M. Rashidi, "Magnetic field effect on unsteady nanofluid flow and heat transfer using Buongiorno model," *J. Magn. Magn. Mater.*, 2016.

[25] M. Hatami, M. Jafaryar, J. Zhou, and D. Jing, "Investigation of engines radiator heat recovery using different shapes of nanoparticles in H<inf>2</inf>O/(CH<inf>2</inf>OH)<inf>2</inf>based nanofluids," *Int. J. Hydrogen Energy*, vol. 42, no. 16, 2017.

[26] O. Torsater, B. Engeset, L. Hendraningrat, and S. Suwarno, "Improved Oil Recovery by Nanofluids Flooding: An Experimental Study," in *SPE Kuwait International Petroleum Conference and Exhibition*, 2012.

[27] L. Levidow, D. Zaccaria, R. Maia, E. Vivas, M. Todorovic, and A. Scardigno, "Improving water-efficient irrigation: Prospects and difficulties of innovative practices," *Agric. Water Manag.*, 2014.

[28] W. Yu, D. M. France, J. L. Routbort, and S. U. S. Choi, "Review and Comparison of Nanofluid Thermal Conductivity and Heat Transfer Enhancements," *Heat Transf. Eng.*, vol. 29, no. 5, pp. 432–460, 2008.

[29] H. T. Zhu, C. Y. Zhang, Y. M. Tang, and J. X. Wang, "Novel Synthesis and Thermal Conductivity of CuO Nanofluid," *J. Phys. Chem. C*, vol. 111, no. 4, pp. 1646–1650, 2007.

[30] E. Magyari and A. Pantokratoras, "Note on the effect of thermal radiation in the linearized Rosseland approximation on the heat transfer characteristics of various boundary layer flows," *Int. Commun. Heat Mass Transf.*, vol. 38, no. 5, pp. 554–556, 2011.

[31] A. E. Kabeel, E. M. S. El-Said, and S. A. Dafea, "A review of magnetic field effects on flow and heat transfer in liquids: Present status and future potential for studies and applications," *Renewable and Sustainable Energy Reviews*. 2015.

[32] M. Khan, M. Y. Malik, and T. Salahuddin, "Heat generation and solar radiation effects on Carreau nanofluid over a stretching sheet with variable thickness: Using coefficients improved by Cash and Carp," *Results Phys.*, 2017.

IOP Conf. Series: Materials Science and Engineering **824** (2020) 012011 doi:10.1088/1757-899X/824/1/012011

Preliminary Studies on Design of Photovoltaic Thermal (PVT) Solar Collector for Roofing System Application.

Zariq Imran Abdul Manap[1], Hanani Abd Wahab[1*], Ong Pauline[1], Al-Emran Ismail[1], Maznan Ismon[1], Zaleha Mohamad[1]
Faculty of Mechanical & Manufacturing Engineering Universiti Tun Hussein Onn Malaysia (UTHM), Batu Pahat, Johor, Malaysia.

E-mail: *nani@uthm.edu.my

Abstract. This study was conducted to develop and study the performance of a Photovoltaic Thermal (PVT) system technology, a system which combines the Photovoltaic (PV) system with a solar thermal (T) system that produces electricity and hot water simultaneously. The objective of this study is to develop a working prototype of the PVT system for roofing system application with water as the cooling agent. A polycrystalline PV cells was used as the solar collector while a transparent rubber tube was attached at the backside of the PV panel with water flowing directly from a water tap source. Two designs were examined for the heat absorber design which was categorized as the parallel and spiral flow design. The system was run at an open space to simulate real life application. A working PVT system was developed; however an electrical efficiency drops 0.3% and 1.07% on the parallel and spiral design respectively when compared to the standard PV system.

1. Introduction

Renewable energy has been the talks of the industry for decades. Based on Malaysia Five Fuel strategy policy, it is targeted for renewable energy to be the source of 11% electricity generation that will subsequently amount to 2080MW of installed capacity by the year 2020. Among the sources of renewable energy available, Solar Photovoltaic (PV) has shown the most significant demand among citizens in Malaysia with a total of 97.63% consumer application received by the Sustainable Energy Development Authority (SEDA) in 2015 as compared with biogas, biomass, small hydro, and geothermal [1].

Photovoltaic (PV) systems introduced in Malaysia in the 1980s, since then the PV cell become one of the best alternative for generating power source. PV systems have many benefits for the user such as, lower maintenance costs [2]. Mostly, PV system manufacturer gives 20 – 25 years warranty with no moving parts, no wear, and tear. Using the PV system can reduce the electricity bills since the PV system just needs the energy from the sun, which is a free and unlimited source to produces the electric power. Also, PV system is very versatile; it can be uses for small or big applications [3].

Photovoltaic thermal (PVT) hybrid solar collector is system that converts solar radiation into thermal and electrical energy. These systems combine a solar cell, which converts sunlight into electricity, with a solar thermal collector, which captures the remaining energy and removes waste heat from the PV module. There have been many studies of PVT systems integrated to building systems focusing on numerical methods, control and design optimisation, production materials, innovative systems and economic analysis [4-12]. Therefore, this study are made to develop and study a hybrid PVT solar collector system in terms of efficiency and application capability by adding a water based cooling system into the panel system.

2. Methodology

2.1. General System Selection

The system selection for the roof PVT module will be a static type module and is suitable for Malaysian standard roofing pattern. The system will be based on an off-grid PVT system. The main components that will be needed for the prototype to work are:

i. Tiled roof
ii. PVT module
iii. Water storage tank
iv. Pumping system
v. Cable and piping

2.2. Type and Size of PV Module

The sizes of the PV module that will be used for the simulation are a $4m^2$ flat plate, unglazed polycrystalline PV cell. This type of cell is used instead of the higher efficient monocrystalline cells due to its lower manageable cost and acceptable working efficiency. Based on the literature review, a polycrystalline cell possesses a working efficiency value of around 12% - 16%. Rekha, Vijayalakshmi, and Natarajan [4] claims that a PV system with a $4m^2$ module area are able to generate as much as 3kWh of electricity per day. This amount will be sufficient as the common current usage of 2 – 4 kWh per day for any normal household of 1-4 persons. The size will also enable the panel to receive 100 liter per day of hot water with temperature of $80^{\circ}C$ [4].

2.3. Collector Properties

The material that was chosen to be used as the collector body is diverse. Using a less expensive material such as steel will not influence any significant change in terms of the absorber thermal and electrical performance even though they have a low thermal conductivity [5, 10]. The design of the collector uses the spiral flow absorber design is as shown in a study by Ibrahim as in Figure 1 configuration of PVT collector design. Whereby, the spiral flow design simulates the best thermal and cell efficiency which is 50.12% and 11.98% respectively [6].

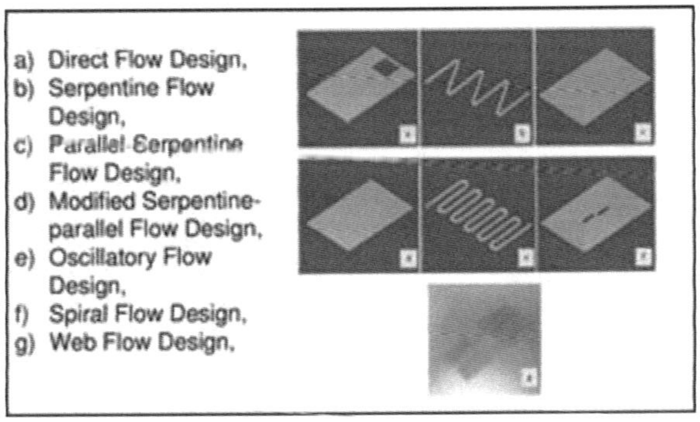

Figure 1. Configuration of absorbers design [6].

2.4. Water Flow Rate

Water flow rate plays a significant role in determining the output temperature of the heated water. If the water flow is too fast, the amount of heat absorption will be lower thus increasing the surface temperature of the PV panel. This means that an optimum rise in water flowrate will affect the rate of heat removal on the collector. The optimum water flow rate that will be used in the simulation will be 20liter/hr. The use of optimum flow rate of 20liter/hr for a PVT system on residential building will increase the PV cells efficiency [7].

2.5. Malaysia Solar Roofing System Application

The majority of buildings in Malaysia for example in Batu Pahat, Johor, has a minimum inclined angle of 30^0. In the aspect of solar installation, the correct angle of installation will provide better sunlight exposure to the PV panel thus creating a more optimum energy production.

2.6. Malaysia Meteorological Condition

The performance of a solar system depends greatly on the meteorological state in a given area. One of the influential factors lies on the strategic geographical location of the nation. Malaysia is located specifically along the South China Sea with coordinates in between of 10 and 70 to the north latitude and 1000 to 1200 on the south longitude. Whereby, Malaysia is capable of receiving sunlight exposure up to 4 to 8 hours a day [8].

2.7. Prototype Modelling and Design

The software that will be used to visualize the design is the SolidWorks. The components used in the design stage are standard commercialized components. This means that there is no major analysis required to be done on them other than determining the suitable tone-down-ratio counterpart to be used on the smaller scaled prototype.

2.8. Component Decomposition Analysis

The decomposition of the components involved in the desired prototype design is as shown in Figure 2. As suggested in the general system selection, the main components of the PVT system consists of the PV module and the thermal absorber, the water tank, and water pump.

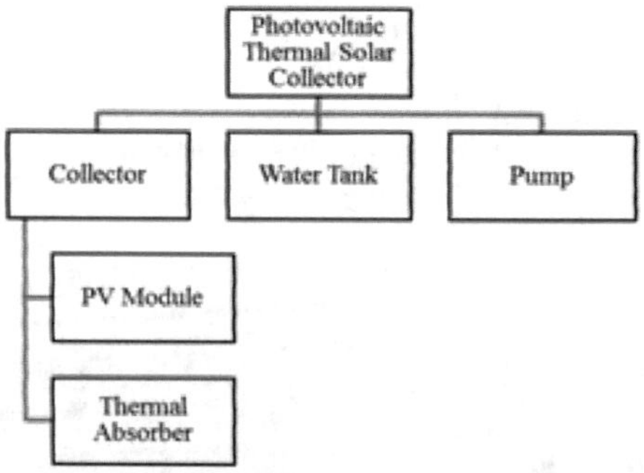

Figure 2. Component decomposition chart for the PVT system

2.9. Function Decomposition Analysis

Figure 3 shows the function of each of the components involved in the PVT system based on the components decomposition chart.

2.10. Function Structure

The function structures explain the functions of the PVT system in the form of a flow chart. Figure 4 shows the overall function of the PVT system where in general, the system received sunlight as an input and produces current and heat as the output.

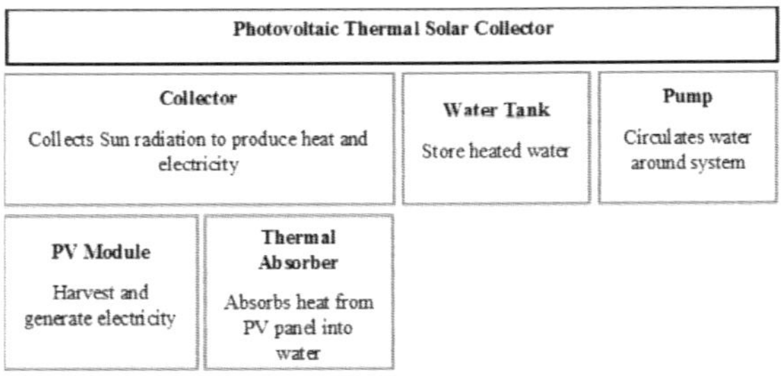

Figure 3. Function decomposition chart for the PVT system

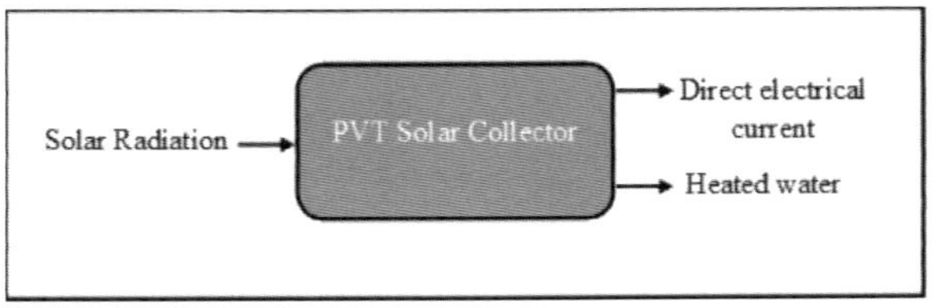

Figure 4. Overall function of the PVT system

Figure 5 illustrate in a more detailed perspective of what is happening inside the system that would generate the output as described in Figure 4. The function flow is shown step by step so that a clear understanding regarding the system function can be obtained. This part is essential in order to have a clean view on what to be expected when designing the prototype so that each and every component are placed and used in their best possible position so that the desired end result will be achieved more efficiently

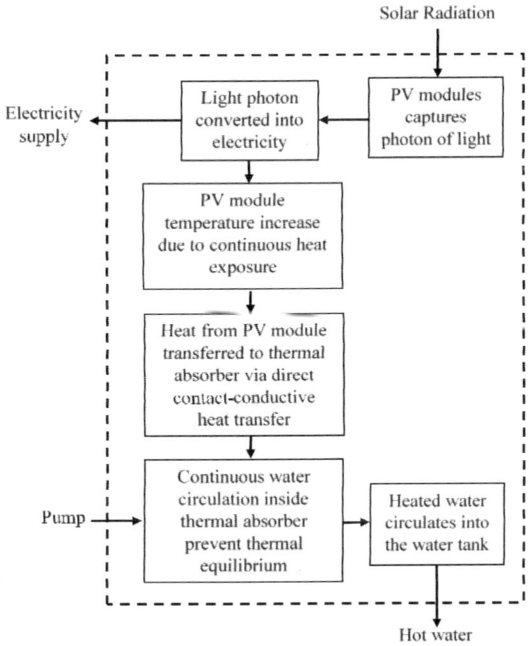

Figure 5. Flow chart of function structure for the PVT system

IOP Conf. Series: Materials Science and Engineering **824** (2020) 012011 doi:10.1088/1757-899X/824/1/012011

3. Result and discussion

3.1. SolidWorks Design

Figure 6 shows the prototype design made on the SolidWorks software which was used as the reference for product fabrication.

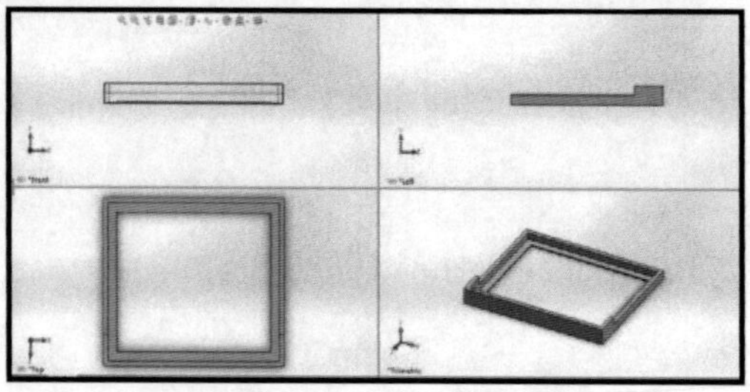

(a) The prototype main frame

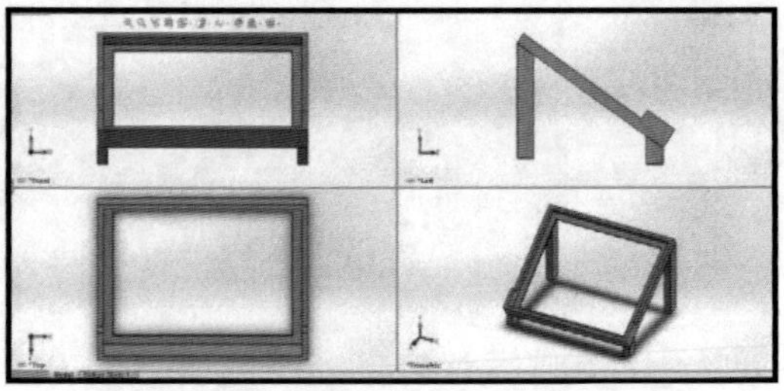

(b) The main frame with a 35°C angle

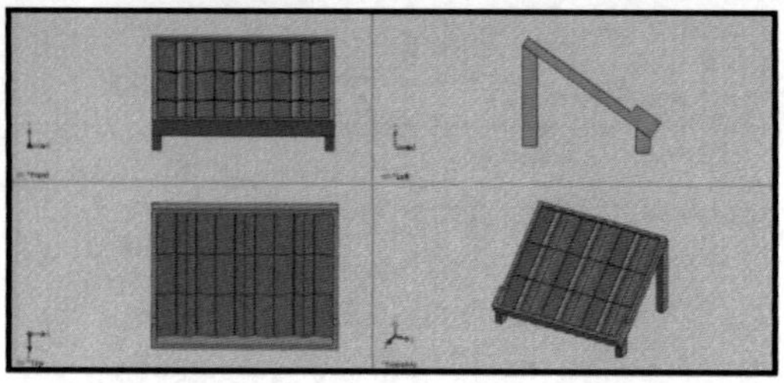

(c) The main frame with roof tile

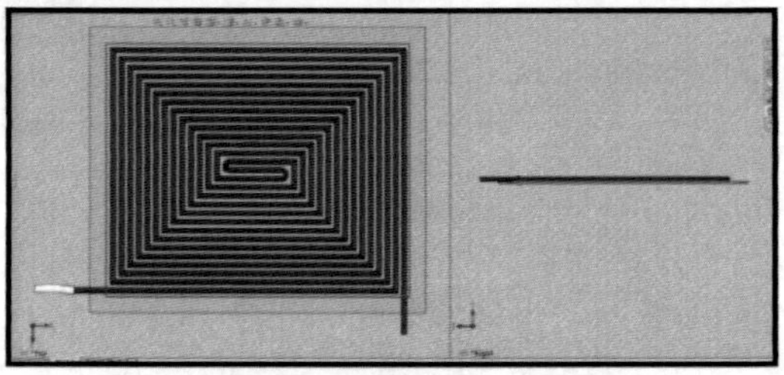

(d) The design for spiral water tube heat absorber

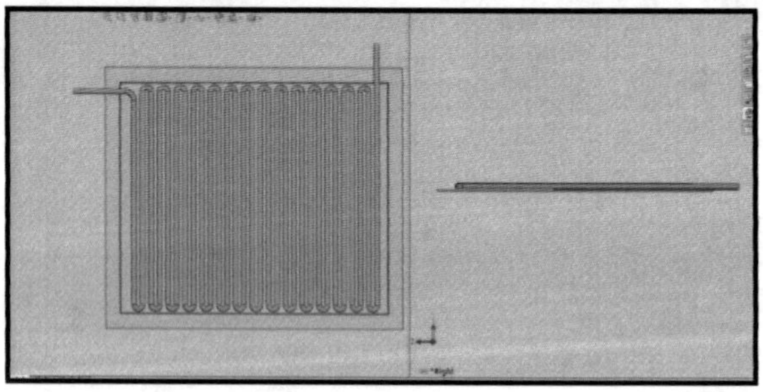

(e) The design for parallel water tube heat absorber

Figure 6. PVT Prototype design using Solidworks

3.2. Fabricated Prototype

Figure 7 shows the experimental setup for PVT system with parallel and spiral absorber design.

(a) Standalone PV setup

(b) PVT setup

(c) PVT-parallel absorber design

(d) PVT-spiral absorber design

Figure 7: The Experimental Setup Design

3.3. Expected Experiment Results

Based on the literature review made at the early stage of this study, it is expected that:

i. The PV panel experience voltage production drop as its surface temperature rise
ii. The PV panel current production is dependant towards the level of solar radiation it receives from the sun which is determined through the weather condition.
iii. The amount of maximum power generated by the panel to be at 10-12 p.m. as the solar radiation peaks at that time.
iv. The amount of power production decreases after 12 p.m. as solar radiation level decreases.

When the system is changed from PV to PVT, it is expected that:

i. The range of voltage production increase compared to the standalone PV system
ii. The panel power generation will increase significantly
iii. The panel surface temperature will be relatively lower than the PV setup.
iv. There is a significant increase in water temperature after it circulated through the PV panel
v. The overall efficiency of the panel in PVT system increases compared to the PV system.

IOP Conf. Series: Materials Science and Engineering **824** (2020) 012011 doi:10.1088/1757-899X/824/1/012011

3.4. Experiment Results and Discussion

3.4.1. Photovoltaic Setup. The photovoltaic setup is the setup in which only the PV panel are placed on the roof frame without the water tube circulation system. This data collected on this setup were made as the benchmark performance that needed to be overcome by the later system designs.

The setup parameters are as follow:

i. Setup : Photovoltaic system
ii. Place : Taman Wira Neighbourhood, Parit Raja
iii. Placement time : 10.00 a.m. to 3.00 p.m.

3.4.2. Photovoltaic Thermal with Parallel Water Tube Design. The photovoltaic thermal with parallel water tube design setup is the improved design in which the PV panel are equipped with a cooling system in a form of water tube. The tube will serve as a heat transfer mechanism which drains the heat from the PV panel into the water inside the tube. As the water circulates, the heat will be transferred away from the panel and hot water will be additionally obtained. The setup parameters and results obtained are as follow:

i. Setup : Photovoltaic-Thermal system with parallel water tube system
ii. Place : Research Centre for Soft Soil (RECESS) UTHM
iii. Placement time : 10.00 a.m. to 3.00 p.m.
iv. Water flowrate : 0.0156 L/s

3.4.3. Photovoltaic Thermal with Spiral Water Tube Design. The photovoltaic thermal with spiral water tube design is the setup that works under the same principle as the previous PVT-spiral tube design. The difference in this system lies on the design of the water tube. Instead of lining the tube parallel, this design sets the tube position in a spiralling manner. The idea of the design is to increase the contact surface area between the tube and the panel, providing a more even and higher potential of heat removal. The setup parameters and results obtained are as follow:

i. Setup : Photovoltaic-Thermal system with parallel water tube system
ii. Place : Taman Wira Neighbourhood, Parit Raja
iii. Placement time : 9.50 a.m. to 3.00 p.m.
iv. Water flowrate : 0.0046 L/s

3.5. Observations

During the experimentation period, it is observed that the PV panel produced the highest and lowest amount of power output at 12 p.m. and 3 p.m. respectively. The weather is seen to be a bit cloudy at 11 p.m. and gradually getting cloudier on 1 p.m. onwards. A good and clear weather condition is observed at 10 p.m. and 12 p.m. while a slight breeze is felt throughout the day. The PV temperature is observed to be hotter than the roof temperature at all time of the day. The upper area of the roof showed a higher temperature range as compared to the lower portion but slowly gained thermal equilibrium as the days goes on.

The PVT panel with parallel design produced the highest and lowest amount of power output at 11 p.m. and 3 p.m. respectively. Contrary to the weather condition for the previous experiment, the weather showed a good sunny condition all day with exception on 12 p.m. which is the only time on the day that is cloudy. The day is also observed to be slightly windy. The PV temperature is also observed to be hotter than the roof temperature at all time of the day. The upper area of the roof showed a higher temperature range as compared to the lower portion but slowly gained even heat distribution at 3 p.m. the PV panel however, is observed to be much hotter as compared to the previous experiment without the water system.

The PVT spiral design panel on the other hand, produced the highest and lowest amount of power output at 1 p.m. and 11 p.m. respectively. This experiment ran under a very good sunny day for all hour. An occasional slight wind was observed at 10 p.m., 11 p.m., and 3 p.m. The PV temperature is also observed to be hotter than the roof temperature at all time of the day. The upper area of the roof showed a higher temperature range as compared to the lower portion but slowly gained even heat distribution at 3 p.m. The PV panel is again, observed to be much hotter as compared to the previous experiment without the water system. It is in fact hotter than the parallel water tube system.

3.6. Data Representation

3.6.1. Ambient Temperature versus Time. Figure 8 shows the relation of ambient temperature versus time for each experiment.

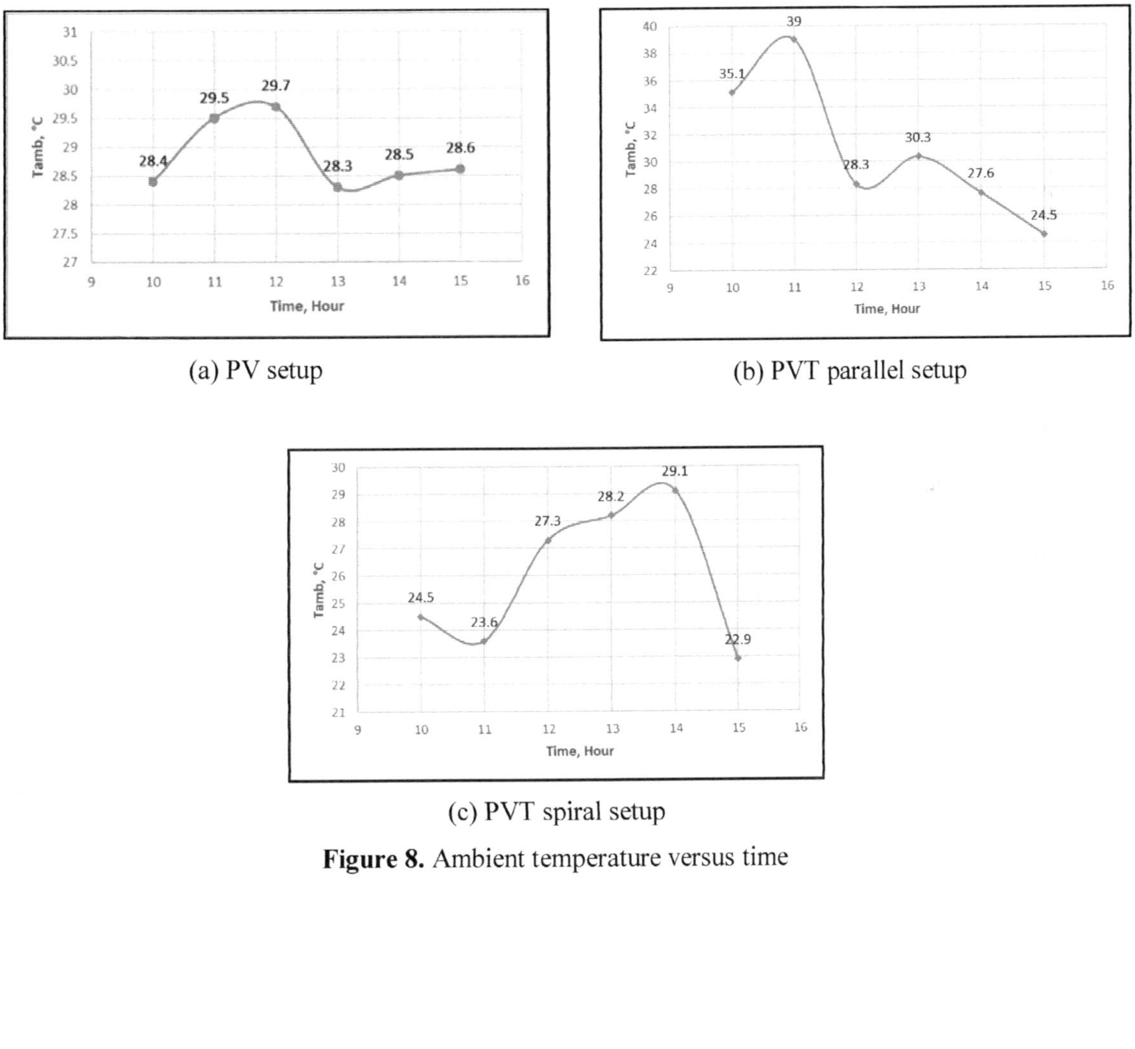

(a) PV setup (b) PVT parallel setup

(c) PVT spiral setup

Figure 8. Ambient temperature versus time

IOP Conf. Series: Materials Science and Engineering **824** (2020) 012011 doi:10.1088/1757-899X/824/1/012011

3.6.2. Roof Temperature versus Time. Figure 9 shows the relation of roof temperature versus time for each experiment.

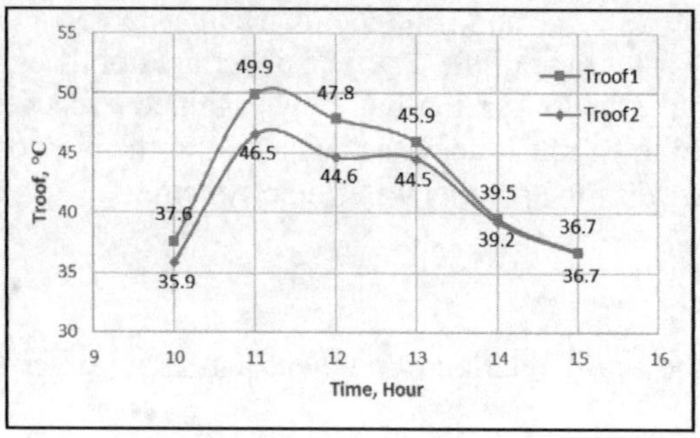

(a) PV setup

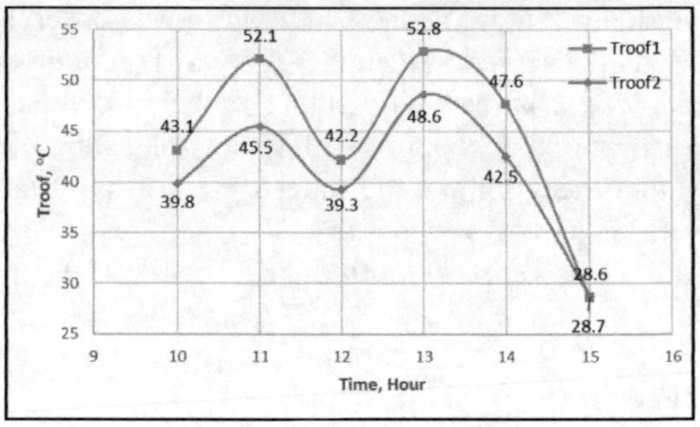

(b) PVT parallel setup

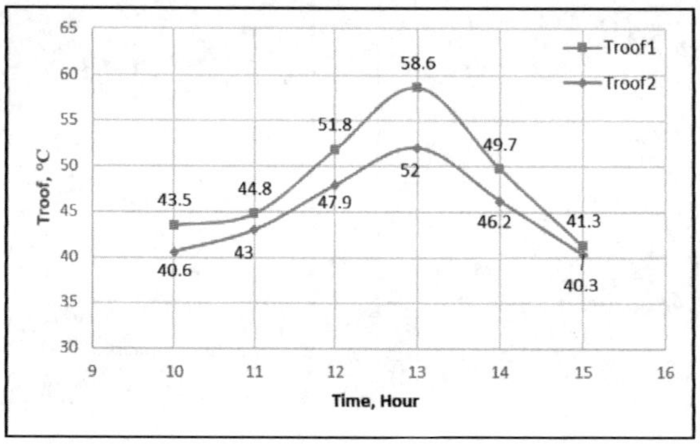

(c) PVT spiral setup

Figure 9. Graph of roof temperature versus time

3.6.3. Panel Temperature versus Time. Figure 10 shows the relation of ambient temperature versus time for each experiment.

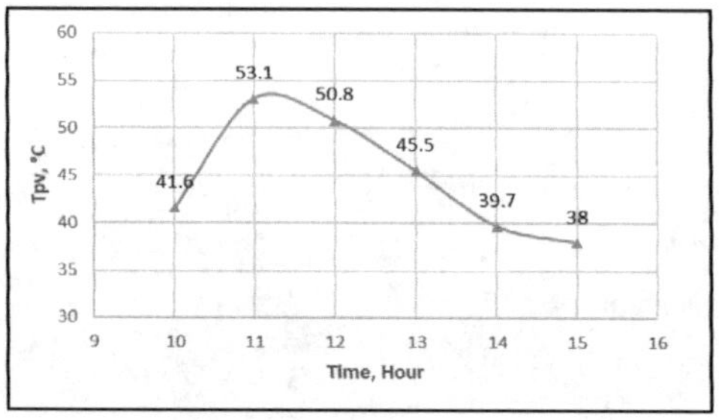

(a) PV setup

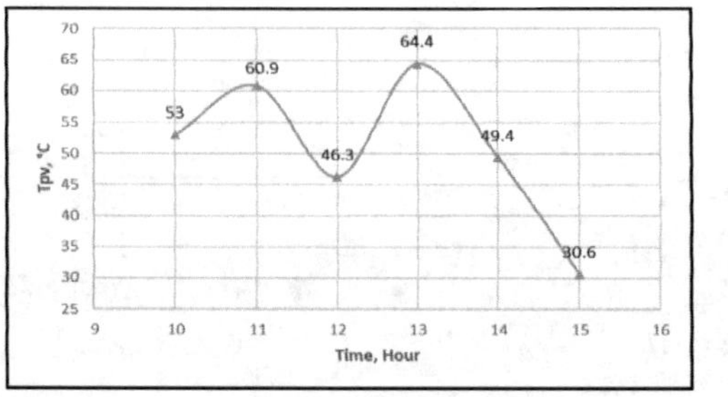

(b) PVT parallel setup

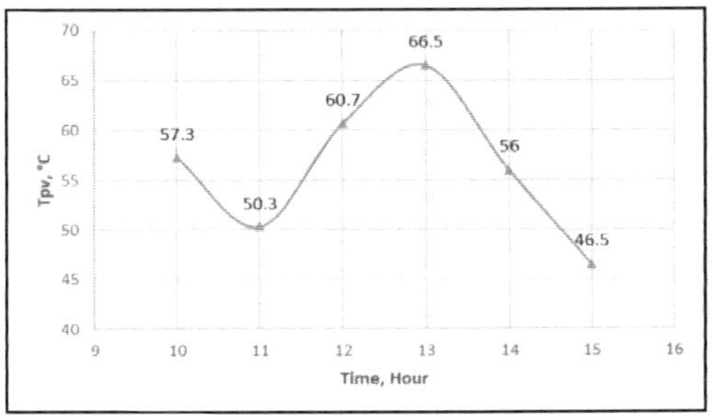

(c) PVT spiral setup

Figure 10. Panel temperature versus time

3.6.4. Voltage versus Time. Figure 11 shows the relation of ambient temperature versus time for each experiment.

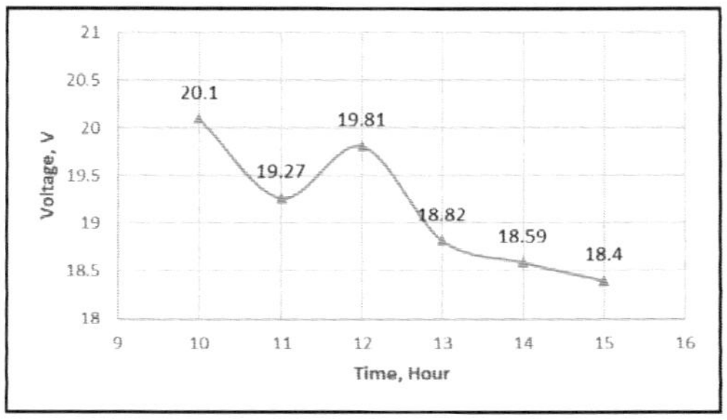

(a) PV setup

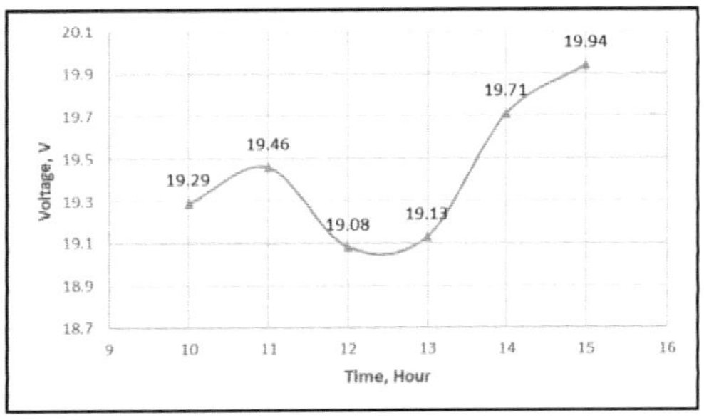

(b) PVT parallel setup

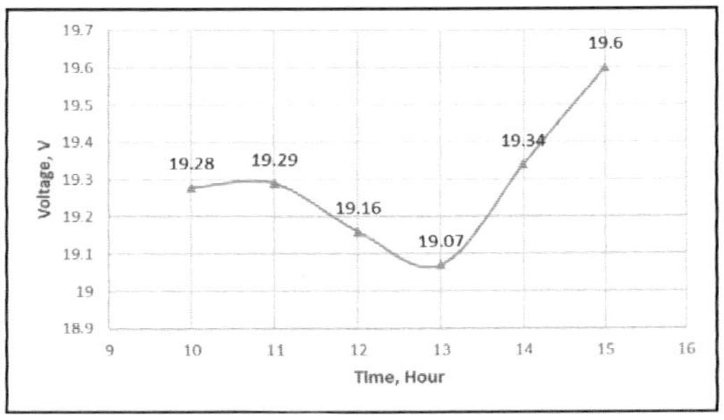

(c) PVT spiral setup

Figure 11. Voltage versus time

3.6.5. Current versus Time. Figure 12 shows the relation of ambient temperature versus time for each experiment.

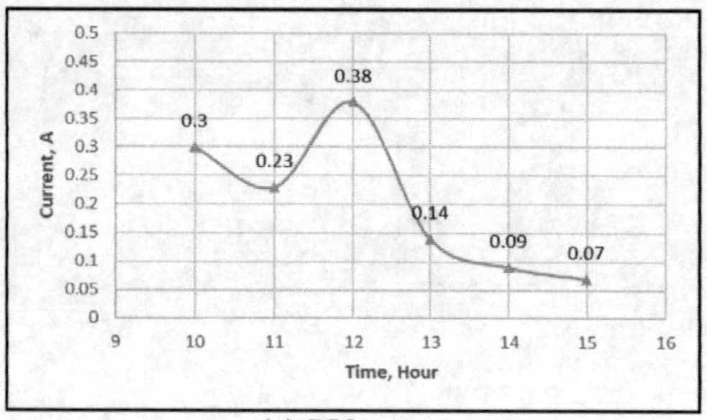

(a) PV setup

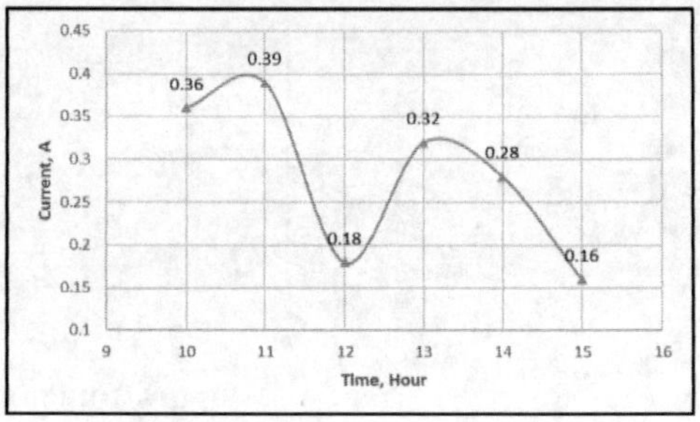

(b) PVT parallel setup

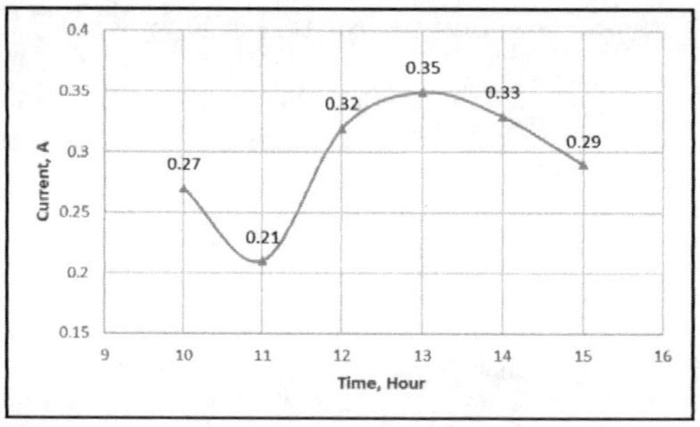

(c) PVT spiral setup

Figure 12. Current versus time

3.6.6. Power versus Time. Figure 13 shows the relation of ambient temperature versus time for each experiment.

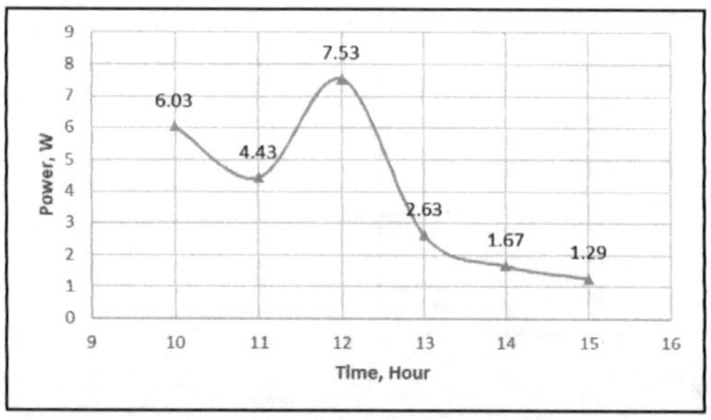

(a) PV setup

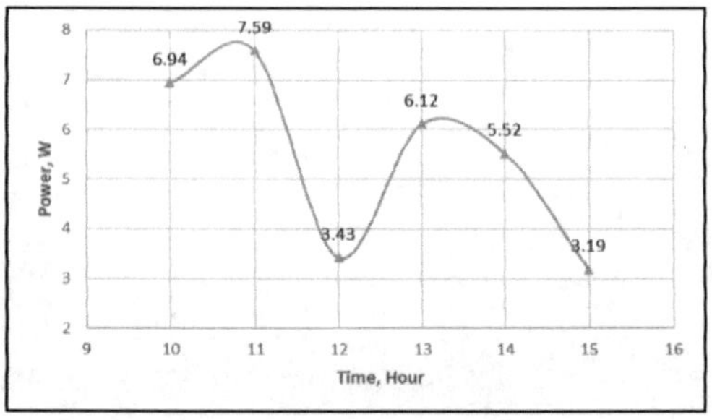

(b) PVT parallel setup

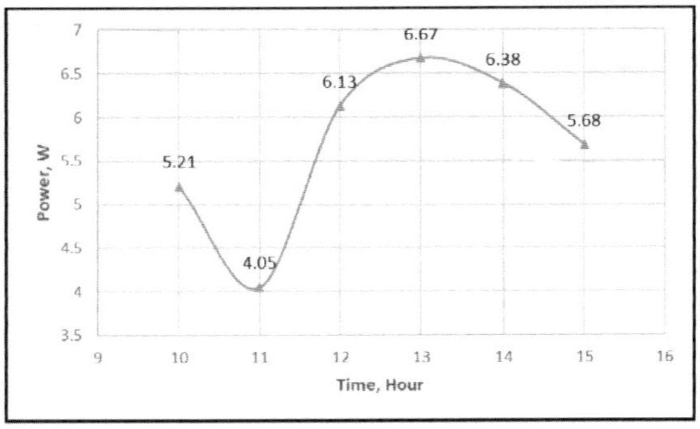

(c) PVT spiral setup

Figure 13. Power versus time

3.6.7. Water Temperature versus Time. Figure 14 shows the relation of ambient temperature versus time for each experiment.

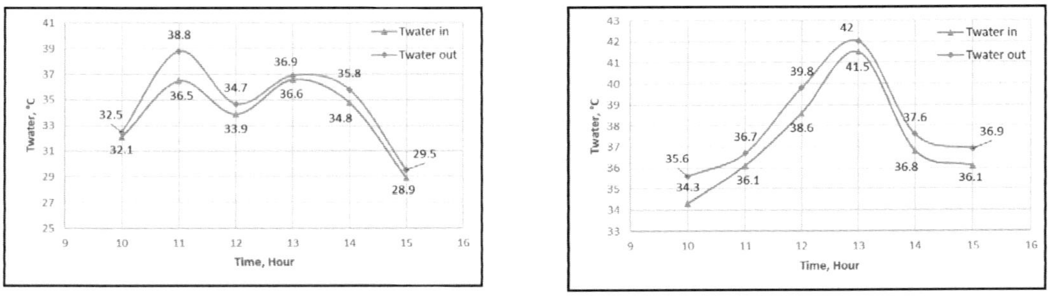

(a) PVT parallel setup (b) PVT spiral setup

Figure 14. Water Temperature versus time for PVT setup

3.7. Data Calculation

3.7.1. General power efficiency. The general power efficiency represents the average power generated by the panel through recorded data. It can be calculated by using the formula below:

$$\text{Efficiency}_{\text{avg}} = \frac{P_{avg}}{P_{max}} \times 100 \tag{1}$$

Where:

P_{avg} = Average power produced under throughout the experiment time
P_{max} = Maximum potential power produced by the panel

3.7.2. Electrical efficiency. Electrical efficiency describes the amount of electrical output generated by the system based on the panel temperature. It can be calculated by using the equation below:

$$\eta_e = \eta_r[1 - \beta(T_c - T_r)] \tag{2}$$

Where:

η_r = PV module efficiency (16.5%)

β = Temperature coefficients (0.0045°C)
T_c = Solar cell temperature
T_r = Ambient temperature

3.7.3. Accumulated heat. The level of heat accumulated on the water system can be calculated by using the equation below:

$$Q_u = \dot{m}C_p(T_\alpha - T_i) \tag{3}$$

Where:
$\dot{m}$ = Mass flow rate of water used
C_p = Heat capacity of collector coolant
T_α = Collector temperature
T_i = Input fluid temperature

3.8. Direct Water-Cooling Method

The result of the previous three experiments showed that both PVT systems failed to demonstrate that water can be used as a working fluid to decrease the PV panel temperature.

This experiment was made in order to justify and prove that water can be used as a working fluid to decrease the PV panel temperature since that the theory were failed to be proven during the previous setup due to various factors such as the tube materials among others. Figure 15 shows the assembled prototype. The experimental setup is as follows:

i. Setup : Photovoltaic-Thermal system with direct water flow above the panel surfaces
ii. Place : Taman Wira neighbourhood, Parit Raja
iii. Placement time : 10.00 a.m. to 12.30 p.m.
iv. Water flowrate : not measured

Figure 15. PVT direct water-cooling setup

3.8.1. Observations and Analysis

The direct water-cooling method recorded a significant cooling effect towards the solar panel. In just one hour of operation, the panel temperature was able to be reduced up to 17.8°C. The PV panel also recorded an increase in voltage production which almost peaks to the maximum. The water heating factor however, can only be seen at the early stage of the experiment.

4. Conclusion

A small scaled prototype of PVT solar with water cooling system were successfully designed and fabricated. The prototype was tested and is able to run as it is intended to. The prototype succeeded in generating electricity and hot water at the same time. Thus, the main objective of this study was successfully achieved. The prototype was also able to demonstrate and prove certain solar panel theories such as the inversely proportional relationship between the solar panel temperature and its voltage production capacity, the dependency of the current production towards the solar radiation which is greatly affected by the weather condition, variables that affect the current and voltage generation will affect the power production, the theory that hot water can be produced and used from the cooling system, and the capability of water to cools down a solar panel and increase it voltage generation. The final theory however, was proven by an additional experiment which is not originally included as one of the prototype designs.

Both prototypes however, failed to demonstrate an increase its performance when compared with the standalone PV panel. By generalizing the result, both PVT design which consists of the spiral and parallel water tube recorded a decrease in electrical efficiency by 0.3% and 1.07% respectively when compared to the PV system. Hot water was able to be produced. There is however, a slight increase in the general power generating efficiency showing that the voltage production was slightly stabilized when the PVT system is used. Reasons for the results obtained are deduced through observation and analytical approach which suggest the study has various flaws regarding material selection, possible data handling procedure, and some uncontrollable factors that affected the data collection.

To conclude, the prototype made for this study was able to run as it is intended to. It was able to produce slightly more stable electricity and produce slightly hot water for additional usage. The performance of the prototype however, does not perfectly demonstrate a more efficient performance when compared with the commercialized PV panel due to several flaws existed during the experimentation period. It can be said that the main objective of this study is achieved in terms of proving the working theories behind the PVT system. The study however, failed to demonstrate the PVT theories through a single working prototype. Therefore, as for further study, it is requires improvement in terms of material selection, experimentation handling and data harvesting methodology.

5. References

[1] SEDA, "Annual Report Sustainable Energy Development Authority Malaysia 2015," 2015.
[2] La Tour, Arnaud de, Matthieu Glachant, and Yann Ménière. 2013. "Predicting the Costs of Photovoltaic Solar Modules in 2020 Using Experience Curve Models." Energy 62: 341–48.
[3] Chua, Shing Chyi, and Tick Hui Oh. 2012. "Solar Energy Outlook in Malaysia." Renewable and Sustainable Energy Reviews 16 (1): 564–74.
[4] Rekha L, V. M. (2013). Photovoltaic thermal hybrid solar system for residential applications. International Journal of Research in Engineering and Technology, 277-283
[5] Anderson, T.N., M. Duke, G.L. Morrison, and J.K. Carson (2009). Performance of a building water integrated photovoltaic/thermal (BIPVT) solar collector. Solar Energy, 83: 445-55.
[6] Adnan Ibrahim, G. L. (2009). Hybrid Photovoltaic Thermal (PV/T) Air and Water Based Solar Collectors Suitable for Building Integrated Applications. American Journal of Environmental Sciences 5 (5), 618-624
[7] Thipjak Nualboonrueng, Pongpith Tuenpusa, Yuki Ueda, and Atsushi Akisawa. (2012). The performance of PV-t systems for residential application in Bangkok. Progress in Photovoltaics: Reearch and Application.
[8] Agung Murti Nugroho, A. (2011). Prelimanary Study of Thermal Environment in Malaysia's Terrace Houses, Journal of Economics and Engineering, 2(1).

IOP Conf. Series: Materials Science and Engineering **824** (2020) 012011 doi:10.1088/1757-899X/824/1/012011

[9] Tang, C.K. (2012). Building Energy Efficiency Technical Guideline for Passive Design. *Chapter 2 – Malaysia's Wheater Data*, 20

[10]H. A. Wahab, M. Duke, J. K. Carson and T. Anderson, "Studies of control strategies for Building Integrated Solar Energy System," 2011 IEEE Conference on Clean Energy and Technology (CET), Kuala Lumpur, 2011, pp. 342-347.

[11] Tidjani, F.S., Chandra, A., Pillay, P. Design of building integrated photovoltaic system to the grid with power quality improvement features for Central African countries (2014) IECON Proceedings (Industrial Electronics Conference), art. no. 7048780, pp. 2023-2029.

[12] Lazos, D., Sproul, A.B., Kay, M. Optimisation of energy management in commercial buildings with weather forecasting inputs: A review (2014) Renewable and Sustainable Energy Reviews, 39, pp. 587-603.

IOP Conf. Series: Materials Science and Engineering **824** (2020) 012012 doi:10.1088/1757-899X/824/1/012012

Multi objective optimization of Indoor UHF RFID Network Based on Gradient - Cuckoo search

Nihad Hasan Talib[1], Khalid Bin Hasnan[2]. Azli BinNawawi[2]. , Adel Elewe[1] , Haslina Binti Abdullah[1]

[1]Faculty of Manufacturing Engineering, University Tun Hussein Onn Malaysia, Batu Pahat, Johor, Malaysia.
[2] Faculty of Engineering Technology University Tun Hussein Onn Malaysia, Pagoh, Johor Malaysia.

Abstract. : Network design, in general is a critical concept due to its effect on efficiency, cost and other significant factors. In recent years, RFID is widely applied for RNP Network design. The large -area network design process requires a significant number of interrogating antennas based on the reader-tag range communication. RFID technology uses a huge number of tags communicate with a small number of the reader, from thus point of view, the challenges of large scale RNP problems include high computational cost due to the time consumption of RFID readers placement error, as a result, these challenges reduce the effectiveness of the RFID system In this study, a model of multi-objective function for RFID reader placement was conducted on various large -area condition to evaluate the impact of network design expansion. A comparative analysis was performed with (GBCS algorithm) Gradient-Based Cuckoo Search. The dataset was performed for the area of 80m x 80m and 150m x 150m. Simulation results exhibited that the performance of (GBCS) Gradient-Based Cuckoo presented a minimum number of deployed readers with maximum RFID tags coverage and was superior in solving large scale RFID-NP problems. Simulation results not only explained that the present algorithm is strong and workable but also showed excellent approximation abilities even in the high scale-area. Consequently, the authors recommend if the area bigger than ($80m^{2}$), there is a need to divide the area into separate regions and deal with each region separately to be sure that the objective function can work efficiently.

1. Introduction

Radio frequency identification system (RFID) as a modern inventory monitoring technology has accomplished significant development in many industrial domains. Many real -applications of an RFID system such as toll collection, warehouse, supply chain management, transportation and manufacturing more and more readers are deployed to provide full coverage of all the RFID tags in the given working area.[1] This case leads to the problem of finding the readability of UHF RFID tags based on the limited domain of the tag-to-reader telecommunication [2] therefor, the RFID network planning aims to improve a set of objectives involving the coverage of RFID tags, cost of installation, avoid interference between RFID readers and economic efficiency, by adjusting control variables of the system, such as the required number of the readers, location of the readers, and the parameters of antenna . In RFID technology, the optimization method was very beneficial in solving problems of high complexity, large search area, and searching ill-structured areas. For these reasons alone, Nature

Inspired techniques have been applied in this field. Optimization methods used to improve the values of chosen parameters of relevant RFID objective functions. The design for optimization begins with the design of objective function. The parameter relationships are formulated by the mathematical model [3]. There are many methods are applied for optimization, with Swarm Intelligence technique is one of the optimization methods that gather the multi- systems. This method includes an efficient technique to solve real-difficult problems [4]. Nowadays, a group of RFID Network design optimization has been developed involving five basic techniques [5]. These techniques are (PSO algorithm) particle swarm optimization. (ABC algorithm) Artificial Bee- Colony,
(BFO algorithm)Bacterial Foraging Optimization, (ACO algorithm) Ant Colony Optimization, and (FA algorithm) Firefly Algorithm. (FA) [6].The cuckoo algorithm was newly developed to use in the wide domain such as engineering optimization and to solve difficult optimization problems [7]. All of these methods were developed to enhance their performance to solve RFID-NP problems. (PSO algorithm)Particle Swarm Optimization was improved by Hasnan et al. [5] using (MC-GPSO algorithm) Multi-Colony Global Particle Swarm Optimization to cluster the working space into regions and apply the algorithm in each region separately. Elewe et al. [8] developed the (FA algorithm) Firefly Algorithm by applying (DBSCAN method) density base technique to scan and investigate the topology of the area. This method called (DBSCAN-FA) represents the result of hybridization between (FA) Firefly algorithm and Density-Based Clustering technique (DBSCAN). The contribution of this method (DBSCAN-FA) is the classification of the RFID tags based on their density and specify the optimum reader numbers and primary reader locations. Finally, the (CS) cuckoo algorithm was developed by Fateen and Bonilla, [9] by utilizing (GBCS algorithm) Gradient-Based Cuckoo Search algorithm. The gradient improves numerical performance and enhances the search method. All these techniques (GBCS and DBSCAN-FA) will be examined in this work to evaluate their performance in the large area RFID Network design. The aim of this work is to evaluate the impact of extended network planning based on the RFID objective function in order to be able to choose the superior method to apply in the real application of large-area IoT field.

2. PNR mathematical model

The mathematical model of RNP in this study was construct based on the circular propagation area of RFID readers. The objective function of the RFID-NP Network Planning was developed to enhace the (QOS) quality of solution. The propagation domain of the reader can be calculated by Friis transformation equation The Friis transformation is applied to find the power received from one antenna (with gain G_T), when transmitted from another antenna (with gain G_R), based on distance r_{max} between antennas, and operating at frequency f as in (1) [10].

$$P_R = P_T \times G_T \times G_R \times \left(\frac{\lambda}{4 \times \pi}\right)^2 \times \frac{1}{{r_{max}}^n} \tag{1}$$

The difference between the power transmitted and received is known as (PL) path loss. From this formula, the PL is higher for higher frequencies and energy transferred at lower frequencies will be highest. The functional parameters that affect the optimization process are identified in Table 1 below [8;10]:

Table 1. UHF-RFID parameters.

Symbols Value	UHF RFID Parameters	
P_R	Power input at tag antenna	[0.1 - 2] watts
T_R	Power Sensitivity- thresholds of reader	-70 dbm
T_T	Power Sensitivity- thresholds of tag	-12 dbm
n	Path loss exponent (free-space)	2
f	Operating Frequency	915 MHZ
G_R	Receiving tag antenna Gain	2
G_T	Receiving reader antenna Gain	2
C	Speed of light	(299792458 m/s)

In Free -Space, the path loss exponent (n) is set at 2, but in a cluttered environments such as (Urban area. Suburban – area, pipes and tunnels, etc.) PL exponent (n) may vary of 1 to 4 [11]. The propagation domain of the RFID reader can be determined in equation (2) – (4) [12].

$$L_{db} = P_T + G_T + G_R + P_R \tag{2}$$

$$L_m = 10^{\frac{L_{db}}{10}} \tag{3}$$

$$r_{max} = \frac{\sqrt{L_m \times \lambda}}{4 \times \pi} \quad where \quad \lambda = \frac{C}{f} = \frac{299792458}{915} \tag{4}$$

The relationship between the propagation range of the reader and tag threshold power sensitivity with tag antenna gain which are specified in Table1. given by equation (4) and plotted in Figure 1.and Figure 2

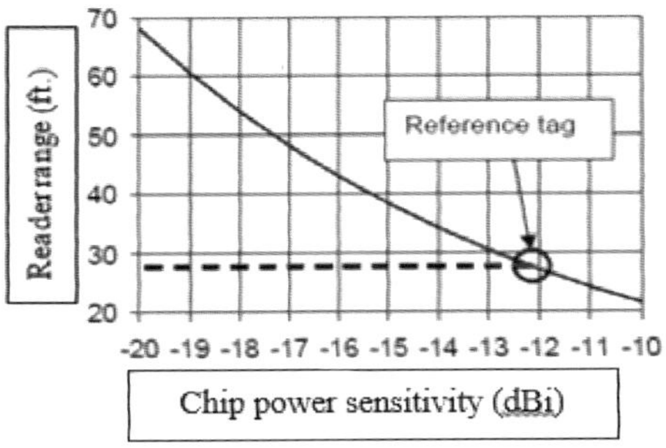

Figure 1. Rrelationship between the propagation range r_{max} of the reader and tag threshold power sensitivity T_T

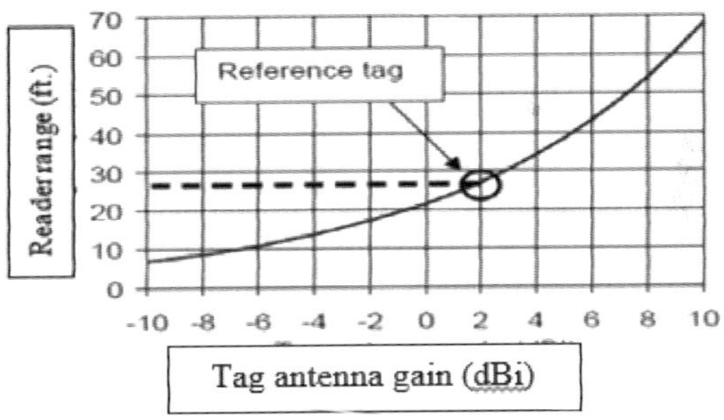

Figure 2. Rrelationship between the propagation range r_{max} of the reader and tag antenna G_R

Therefore, by using equation (4) we can note from figure (1) and figure (2) the propagation area is dependent on RFID tag antenna gain and tag threshold power sensitivity Accordingly the maximum propagation area (r_{max}) not exceed 29ft when tag threshold power sensitivity is set at (-12dbm) theoretically. presently, the maximum range of UHF RFID technology is restricted by tag threshold power sensitivity. However, when the tag threshold power sensitivity is high, such as in the case of semi-passive tags, the range becomes restricted by the reader sensitivity. Note that tag range in UHF RFID can also be affected by various kinds of overlapping, like reader-to-tag, tag-to-tag, and reader -

IOP Conf. Series: Materials Science and Engineering **824** (2020) 012012 doi:10.1088/1757-899X/824/1/012012

to reader. The propagation range is calculated in the above formula in equation (4) Is subject to the boundary conditions [6]

$$r_{max} \geq r_{td} \quad (5)$$

The distance between tag and reader in working space can be calculated by the following expression:

$$r_{td} = \sqrt{((x - x_t)^2 - (y - y_t)^2)} \quad (6)$$

Where y, x are Coordinate of reader and x_t, y_t are Coordinate of tag.

To calculate the optimal number of RFID readers and its location first the next formula is utilized [11]

$$COV_i = \sum_{t=RS}^{max} (r_{max} - r_{td}) \quad (7)$$

The tags coverage rate(cov) in the specified area, which represents a quite significant and important objective function of RFID technology is determined as:

$$f(cov) = \sum_{t=TS}^{MAX} \frac{cov_i}{Nt} = \frac{detected\ tag}{total\ tags} \quad (8)$$

The overlapping between RFID readers can be reduced by the next mathematical equation [12-14]

$$I_{min} = \sum_{i=1}^{N-1} \sum_{j=i+1}^{N} (r_i + r_j) - (dis(R_i, R_j) \quad (9)$$

Where $(ri, rj) are$ Interference rang and (Ri, Rj) are Position of reader

$$f(I) = \frac{\sum_{ij} cov_i \cap cov_j}{Nt} \quad (10)$$

Based on the presented set of formulas, a RFID-network design performance comparison was made in various RNP cases in order to examine the algorithm accuracy in large - scale situations

3. Result and discussion

The Gradient Based Cuckoo Algorithm operates by creating cuckoo species based on the parameters specified in table 2. The main idea is to lay the cuckoo eggs as a host. In other words, the search will apply the RFID objective function as host. Each selected egg represents a new solution in the algorithm (i.e. new reader positions based on the maximum tags coverage). The best solution will be presented by the best egg by throwing out the weak solutions. The main formulas are shown in equation (11) to equation (15) below:

Table 2. The GBCS parameters.

Symbols	Parameters	Symbols	Parameters
$(\oplus)$	The multiplications of entry-wise	p_a	The nest fraction
x_{kt} *and* x_{it}	Two different solutions	α	Random number intended by the uniform distribution
d_{fi}	The objective function gradient	$\vartheta_f / \vartheta_i$	The process variables

The new solution of xi (t+1) for the *ith* cuckoo will use the following Lévy flight equation (11) below:

$$x_i(t+1) = x_i(t) + \alpha(\oplus)levy(\lambda) \quad (11)$$

The Lévy flight in above equation is considered as step-lengths based on the following probability distribution in equation. (12) Below:

$$levy\ \ u = t - 1 \qquad 1 < \lambda < 3 \quad (12)$$

The modification presented by Petriciolet in 2014 applied the random walk based on the fraction (1-pa) of the replaced nests, (1-pa) of the nest selected at random is abandoned and changed by new ones at new positions via local random walks. The local random walk can be written as equation (13) below:

$$x_i^{t+1} = x_i^t + \alpha\left(x_i^t - x_i^k\right) \quad (13)$$

To modify the present algorithm, the researcher resaved the randomness of the magnitude of the step. However, direction is calculated based on the gradient sign of the objective function. When the gradient is negative, the direction of step will be positive. If the gradient is positive, the direction of step will be negative. Based on the present sequence, new nests will be generated randomly from the worst nests but in the direction of the minimum number of old nests. The simple modification of the original cuckoo search algorithm was done to enhance the numerical performance of stochastic optimization methods particularly the precision and quality of global optimum solution. Thus, equation (11) is replaced by equation (14);

$$x_i^{t+1} = x_i^t + step_i \otimes sign\left(-\frac{step_i}{df_i}\right) \quad (14)$$

$$step_i = \alpha\left(x_j^t - x_k^t\right) \quad (15)$$

Where sign function involves the sign of its argument and df_i is the objective function gradient at each variable, which is ∂f/∂i. The results of this method show an improvement in fitness function effect based on the coverage efficiency and also specified the optimum installation reader positions [9; 11].

GBCS operates in the following steps:

The step by step operating procedure of gradient cuckoo algorithm based on the specified parameters of RFID is explain as follows:

1- Randomly deploy the number of nests in working space, the nests must be equal to number of initial RFID readers.

2- Evaluate the fitness value of each RFID reader based on the (OF) objective functions used in the equations (1 to 10), then record the "Best Solution" of each reader.
3- Update the gradient position of all readers by using equation (14).
4- Evaluate the fitness value of each RFID reader then compare the results with the previous best fitness position.
5- If the achieved fitness value so far is global best with optimal reader position and the iteration has to be completed, then go to the end.

4. Simulation results

The simulation results were conducted with the selected Gradient Based Cuckoo Algorithm presented in Figure 4 and Figure 3. Figure 3 presents the scenarios of ($80m^2$) working space. The RFID tags distribution were 100, 50 and 30. Figure 4 presents the scenarios of ($150m^2$) working space. The tags distribution were also 50, and 30. All the results were based on the specific parameters of UHF RFID Network objective function presented in previous segment. Two kinds of Network Planning instance were conducted in this work, the clustered kind which is indicated by C100, C50, C30, and the random kind which is indicated by R100, R50, and R30. The run was conducted by MATLAB program 2018 with 12000 iterations and 40 independent. The readers denoted by red star as a coordinate center and the RFID tags are denoted by plus sign in blue color. With red dashed line circle for indicate interrogation range of RFID reader. The plotted results observed the RFID readers were distributed in two different areas in which id detected and covered the tags based on different probability. The results are compared with DB-FA algorithm results presented by Adel Elewe et al., 2017, [13-15]. He simulated the 30, 50 and 100 tags in the area of ($80m^2$) and obtained the results presented in Table 3:

Table-3. Comparative Analysis of the Best Results of (80m X 80m) working space.

Tags Number	Algorithm	Reader Number	Interference	Coverage
C30	GBCS	5	0	100 %
C30	DB-FA	5	0	100 %
C50	GBCS	5	0	100 %
C50	DB-FA	5	0	100 %
C100	GBCS	5	0	100 %
C100	DB-FA	6	0	100 %
R30	GBCS	6	0.005	92.5 %
R30	DB-FA	7	0.006	92 %
R50	GBCS	7	0.009	94 %
R50	DB-FA	8	0.002	98 %
R100	GBCS	6	0.009	94.5 %
R100	DB-FA	8	0.007	97 %

The results observe an enhancement in the objective function in all cases when using GBCS algorithm. And Comparative analysis observes the weakness in tag coverage in cases R50 and R100.in cases R50 and R100 the GBCS algorithm was able to achieve 94% and 94.5% coverage respectively, while DB-FA covered 98% and 97% .This is due to the DB-FA algorithm is working effectively in random data because using either uniform distributions or Gaussian distributions to generate novel explorative moves. As well as, they utilize exhaustive tuning; these two points make DB-FA algorithm working high efficiently with random data only.

The same scenario has been used on the larger region in order to explore the impact of the large - area such as warehouse .railway stations. The selected algorithm is GBC due to it presents good ability with large -scale cases. The results are shown in Table 4. It is seen that when the working space is enlarged, the domain of RFID tags coverage is reduced Another observation is the impact of random tags distribution. Which weaken the capability of the GBCS algorithm to discover the tags as well as specify optimum RFID reader to each group of tags. The major reasons for that are the limited propagation range against the larger space between RFID tags and a large number of possibilities in the large area induce many obstacles to reach the best solution as observes that from the time of each run. The times of running in the second instances were five times the first instances in this paper. thus, as a solution to achieve good results for RFID Network design when it is needed to apply the algorithm in this size of working space, there is a necessity to divide the working space into small regions to deal with each region separately to be sure that the RFID objective function can work better. Therefore the authors recommend, that for areas larger than 80m x 80m, to cluster the area into regions and apply the algorithm in each region separately, a solution that will gives better results. Also the run was conducted by MATLAB 2018 with 12000 iterations. The number of iteration is few if compared with state-of-the-arts swarm intelligence iterations. Therefor few numbers of iterations mean reduce (time, effort and cost and fast convergence to find global best solutions) also the GBCS achieved a minimum number of the RFID reader in all cases also means (Reduce purchase and installation costs .Therefor GBCS improves the performance in terms of convergence rate, computation time, and cost.

Table 4. Comparative Results of (150m x 150m) working space.

Tags Number	**Algorithm**	**Reader Number**	**Interference**	**Coverage**
C30	GBCS	6	0.01	85.3%
C50	GBCS	8	00.1	76%
R30	GBCS	6	0	71%
R50	GBCS	7	0	67%

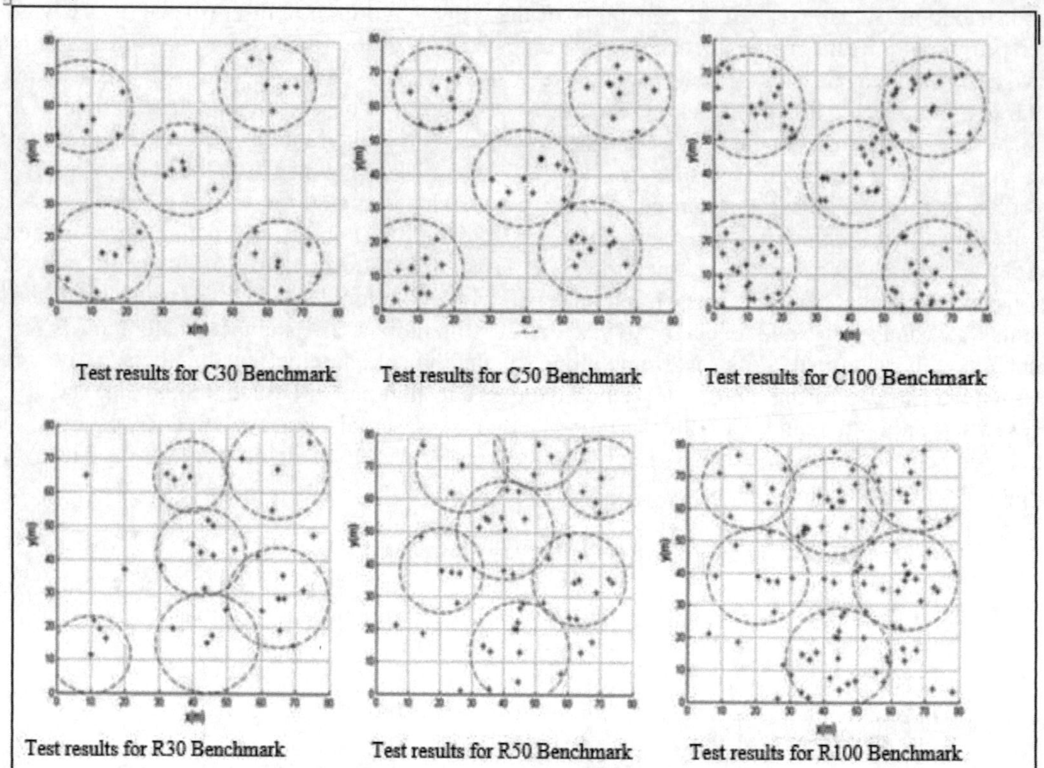

Figure.3 Simulation results of (80×80) m2 scenarios
The clustered kind denoted by C100, C50, C30, and random kind denoted by R100, R50, and R30.

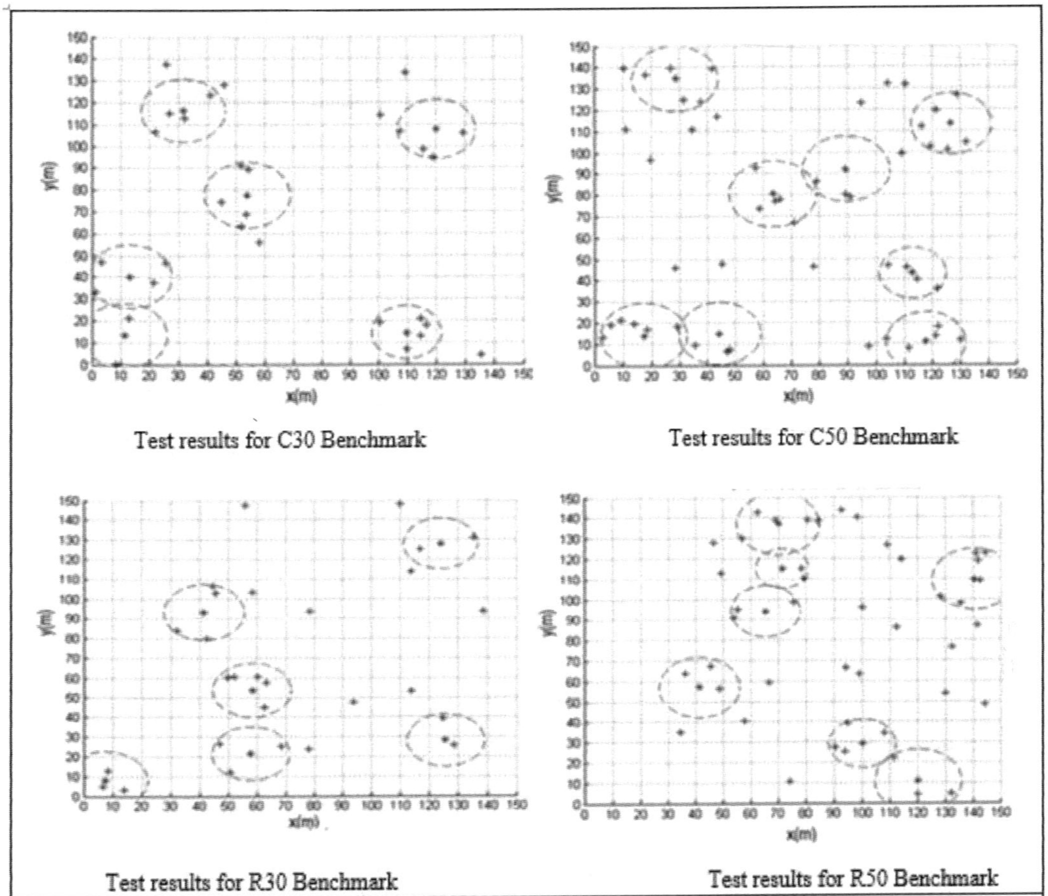

Figure.4 Simulation results of (150×150) scenarios
The clustered type denoted by C50 , C30, and random kind denoted by R50 and R30

5. Conclusion

A survey into best nature inspire algorithms in large scale Radio frequency identification technology Network design is crucial when it is necessary to apply them in real application such as supply chain management or controlling train stations by using RFID system. In this paper, we have compared two state-of-the-arts swarm intelligence based multi objective algorithms, namely, (GBCS) and (DB-FA) for that the researchers tested and compared these methods in large scale area $(80\times 80m^2)$, then enlarged the area to $(150\times 150)m^2$ to investigate the response of the used algorithm. The aim of the present work is to measure the weakness of applying the search algorithm in abnormally large area and select the best method to use in real case of large scale IoT field. The results show that GBCS was the best algorithm for detecting large number of tags and minimum number of readers as well as minimization of reader interference. The benchmark tests also observe that GBCS improves the performacne in terms of convergence rate, computation time, and cost. Results guide the researchers to apply the same algorithm with larger area based on the same condition of dataset. The results demonstrate weakness in the algorithm operation due to large distance between the tag positions. The main reason for that is the limited propagation range against larger distance between tags Therefore, the authors recommend that if the area is bigger than **80**m x **80**m, to cluster the area into regions and apply the algorithm in each region separately. This solution will give better results.

IOP Conf. Series: Materials Science and Engineering **824** (2020) 012012 doi:10.1088/1757-899X/824/1/012012

5. References

[1] Chang TH, Hsu SC, Wang TC (2013) A proposed model for measuring the aggregative risk degree of implementing an RFID digital campus system with the consistent fuzzy preference relations. Appl Math Model 37(5):2605–2622

[2] Chen HN, Zhu YL, Hu KY, Ku T (2011) RFID network planning using a multi-swarm optimizer. J Netw Comput Appl 34(3):888–901

[3] P. Civicioglu, & E. Besdok,. A conceptual comparison of the Cuckoo-search, particle swarm optimization, differential evolution and artificial bee colony algorithms. (2013) Artificial Intelligence Review, 39(4), 315–346

[4] Ali N, Othman M. A., Husain M. N. and M. M. H. 2014.A Review of Firefly Algorithm. ARPN Journal of Engineering and Applied Sciences. 9(10): 1732-1736

[5] Gong Y.J., Shen M., Zhang J., Kaynak O., Chen W.N. and Zhan Z.H. 2012. Optimizing RFID network planning by using a particle swarm optimization algorithm with redundant reader elimination. IEEE Transactions on Industrial Informatics. 8(4): 900-912

[6] A. M. Elewe, K. Bin Hasnan, and A. Bin Nawawi.' Optimization of RFID network planning using MDB-FA method'. (2017)

[7] Jr, I. F., & Yang, X. Cuckoo Search: A Brief Literature Review, 1–13. 2014

[8] A. M. Elewe, K. Bin Hasnan, and A. Bin Nawawi.' 'Hybridized Firefly Algorithm for Multi-Objective Radio Frequency Identification (RFID) Network Planning', 12(3), pp. 834–840.2017

[9] S. Fateen, & A. Bonilla-Petriciolet. Gradient-based cuckoo search for global optimization. Mathematical Problems in Engineering, 2014

[10] Nihad HasanTalib, et al., 2019. Comparative Evaluation of the Gradient -Based Cuckoo Search (GBCS) and (MC-GPSO) Technologies for Optimal RFID Network Planning. International Journal of Mechanical Engineering and Technology (IJMET).Volume:10, Issue: 1, Pages: 677-689

[11] N Hasan Talib, K bin Hasnan, A bin Nawawi, H Binti Abdullah.' Real - Time Transportation Monitoring Using an Integrated RFID - GIS Scheme'. International Journal of Mechanical Volume 9, Issue 13, December (ICME2018);

[12] Hasnan, Khalid Bin, Nihad H. Talib and Nawawi. "Analysis of gradient-based cuckoo search for the large scale optimal rfid network planning." (2019).

[13] bin Hasnan, Khalid, et al. "Comparative evaluation of firefly algorithm and MC-GPSO for optimal RFID Network Planning." Information Technology (ICIT), 2017 8th International Conference on. IEEE, 2017.

[14] N. Bacanin and M. Tuba, and Strumberger, I. (2015) 'RFID Network Planning by ABC Algorithm Hybridized with Heuristic for Initial Number and Locations of Readers'

IOP Conf. Series: Materials Science and Engineering **824** (2020) 012012 doi:10.1088/1757-899X/824/1/012012

[15] A. M. Elewe, K. Bin Hasnan, and A. Bin Nawawi. (Hybridized Firefly Algorithm for Multi - Objective Radiofrequency Identification (RFID) Network Planning) ARPN Journal of Engineering and Applied Sciences Vol. 12, No. 3, Februaury 2017

Verification of Mathematical Model for Upper Limb Spasticity with Clinical Data

J Yee[1], C Y Low[1,*], P Ong[1], W S Soh[1], F A Hanapiah[2], N A Che Zakaria[3], S von Enzberg[4], L Asmar[4] and R Dumitrescu[4]

[1] Faculty of Mechanical and Manufacturing Engineering, Universiti Tun Hussein Onn Malaysia, 86400 Parit Raja, Johor, Malaysia.
[2] Faculty of Medicine, Universiti Teknologi MARA, 47000 Sungai Buloh, Selangor, Malaysia.
[3] Faculty of Mechanical Engineering, Universiti Teknologi MARA, 40450 Shah Alam, Selangor, Malaysia.
[4] Fraunhofer Institute for Mechatronic Systems Design, 33102 Paderborn, Germany.

E-mail: cylow@uthm.edu.my

Abstract. Healthcare providers in the field of physical and rehabilitation medicine play a vital role to help patients suffering spasticity readapting themselves to their normal daily activities. Mathematical modeling of spasticity has the potential to avoid the issue of variability in the assessment of spasticity using the Modified Ashworth scale (MAS). In this work, an existing mathematical model for upper limb spasticity is verified using clinical data sets of upper limb spasticity collected in Malaysia at the level of MAS 1+. The data set consists of torque values measured at each elbow angle as the elbow extends from a full flexion position to a full extension position during slow and fast stretch of the forearm. The aim is to find out the capability of the mathematical model and lay a foundation for the future work on data-driven modeling of upper limb spasticity based on the Modified Ashworth Scale.

1. Introduction

Spasticity is a velocity-dependent reflex contraction of muscle which will result in the inability in muscle control of human body[1]. Upper limb spasticity (ULS) is a more specific term for the manifestation of spasticity on one's upper limb or arm. ULS will affect the normal movement of arm of the sufferers, as the muscles of one's arm will be stiff and difficult to control. ULS is usually seen in patients or victims who experienced stroke, cerebral palsy (CP) or traumatic brain injury due to the damage on the central nervous system[2]. A patient could slowly restore the normal functionality of spastic arm by undergoing proper rehabilitation. However, rehabilitation physicians and therapists have to be well-trained before they are able to provide top-notch healthcare services to the ULS sufferers in order to avoid inflicting injury to the patients[3].

In the current practice, the trainees and students in the field of physical and rehabilitation medicine may at first learn and feel the signs of ULS through role-play sessions with the other peers before proceeding to clinical training to reduce the possible injury as a result of low proficiency. This is due

[1] To whom any correspondence should be addressed.

to the higher awareness in societal responsibility for a lower tolerance of margin for medical errors[4]. Thus, new trainees in the field are not expected to have direct physical encounter with a patient before undergoing clinical training. However, without direct feeling and sensation of a spastic arm, the trainees would not be able to fully grasp the signs and essence of ULS.

Simulation for the medical education appears to be one of the workable solutions to be better equipped in understanding ULS besides directly engaging with patients. The types of simulation that can be employed in medical education range from low-technology simulation to a realistic patient simulator[5], with different complexity, fidelity, and cost. Simulation for ULS is still a less explored area in the research study. In the field of rehabilitation, there are recently a few haptic devices developed for training purpose. The haptic devices used for ULS rehabilitation training are commonly known as upper limb disorder simulator or ULS part-task trainer and can imitate the signs of spasticity in upper limb. However, these devices have their own limitation[6].

Several ULS part-task trainers currently existing in the academic circle include a haptic device for elbow joint spasticity simulation of a child's arm[7], the Haptic Elbow Spasticity Simulator (HESS)[1,8–10], and ULS part-task trainer named BITA[2,11–16]. In order to produce a ULS simulator with a higher fidelity, the HESS research team has developed a set of mathematical equations to describe the relationship between the movement of elbow angle and the resistance produced by the spastic arm[9]. This paper is about verification of the mathematical model based on the clinical data collected from Malaysian patients with ULS. A further analysis on the real and simulated clinical data is conducted to compare and verify the functionality of the developed mathematical equation.

2. Current program

If you don't wish to use the Word template provided, please set the margins of your Word document as follows The This paper will be mainly built upon the foundation of the previously developed ULS part-task trainer named BITA. Currently, BITA is using a primitive program in generating the corresponding torque for the different angle of elbow according to the speed of the stretches, namely slow stretch and fast stretch. When the robotic arm is passively extended slowly which is still lower than a preset angular velocity threshold, the robotic arm would behave according to the torque value assigned to the respective angle in the "slow stretch torque value set". However, if the robotic arm is passively extended beyond a range of degree within a preset time value, the robotic arm enters the fast stretch mode, The function will behave as shown in Figure 1.

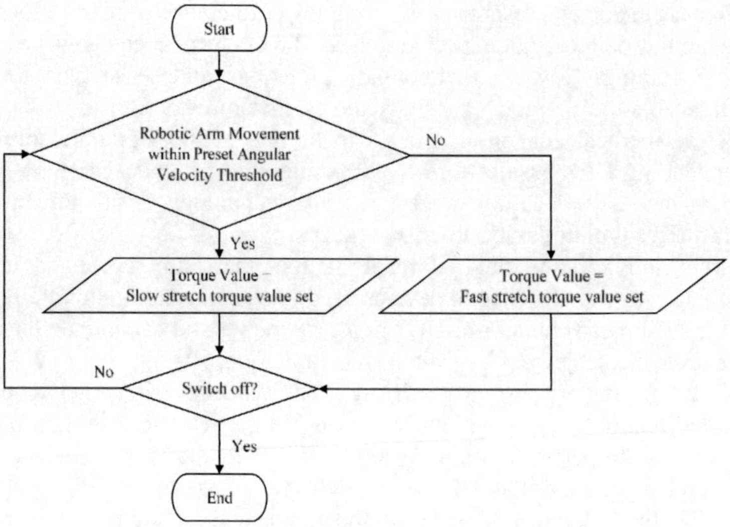

Figure 1. The function loop of the current computer program in BITA

3. Methodology

3.1 Clinical data collection

Sessions of clinical data collection have been conducted in two Malaysian hospitals/specialist centre. The clinical data collection is approved by the ethics committee of the Ministry of Health Malaysia and Universiti Teknologi MARA (UiTM). The clinical data collection system was designed and assembled by the research team of UiTM with the clinical data collection was supervised and conducted by a Consultant Rehabilitation Physician from the Faculty of Medicine of UiTM.

Prior to the clinical data collection phase, the measurement system and equipment were developed. The clinical data collection system consists of two sensors and the whole data acquisition system. The main equipment involved in this measurement system includes a goniometer, USB data-acquisition (DAQ) interface, muscle strength meter (MSM), and electrode amplifier. The specification of the equipment used is summarised in Table 1.

Table 1. Equipment list for clinical data collection system

Equipment	Ref.	Brand	Specifications	
Goniometer (Vernier GNM-BTA)	[17]	Vernier	Range	0-340° (±170°)
			Accuracy	±3°, ±1° with calibration
			Resolution 13-bit(SensorDAQ)	0.06[°]
			Stored calibration values:	Slope: 90[°/V]
				Intercept : -45[°]
USB DAQ interface (Vernier SensorDAQ)	[18]	National Instrument & Vernier	Max sampling rate	48,000 samples/second
			I/O Connectors	3 Analog, 1 Digital
			Resolution	32 bits (16 bits for output)
			Operating system	Windows 2000/XP
Electrode amplifier (Vernier EA-	[19]	Vernier	Power	7 [mA] @ 5V DC
			Input range	-450 to +1100 [mV]
			Impedance	100 [MΩ]

IOP Conf. Series: Materials Science and Engineering **824** (2020) 012013 doi:10.1088/1757-899X/824/1/012013

BTA)	Stored calibration (mV)	Slope : 466.875 Intercept : -559.793

All the equipment listed has different functions in the data collection system. The main data to be collected and analysed for upper limb spasticity are the angle of elbow and the corresponding torque. The function of goniometer is to track and record all the instances of different angle of the elbow to be paralleled with the other parameters. Secondly, the muscle strength meter is used to sense, record, and transmit the data of the force exerted by the physician upon the patient. This is only valid at the fact that the force exerted by the physician is equal to the resistance force of the patients' spastic upper limb. The existence of the DAQ interface in the system is for the connection between Vernier goniometer and a computer while the electrode amplifier is mainly acting as a converter from Bayonet Neill–Concelman (BNC) to British Telecom Connector (BTA) which is able to amplify the voltage into a monitorable range for the interface.

Apart from the main equipment and system, a video recording system is used to record the process of collecting the data for further reviewing purpose in any case when the video is required as a proof.

The clinical measurement process is divided into three different phases: the pre-assessment phase, assessment phase, and post-assessment phase. The pre-assessment phase is the preparation and setting up of equipment including all the sensors, testers, and the recording equipment. The assessment phase is when the therapist engages with the patient for the slow stretch and fast stretch followed by the evaluation of severity level according to the Modified Ashworth Scale (MAS). The post-assessment phase is about analysing the data to extract the valuable information for training. The overall process flow is summarised in Figure 2.

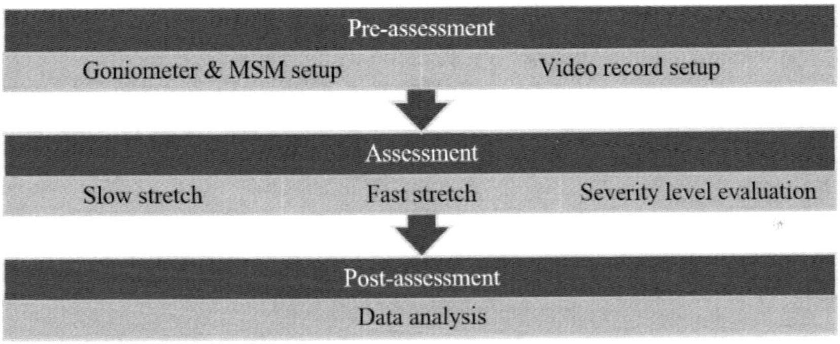

Figure 2. Flow chart for clinical data collection and assessment

The pre-assessment phase is the preparation of all the equipment for the assessment including the whole measurement system of goniometer and muscle strength meter, together with the whole video recording setup. The assessment process is as such: the therapist will conduct 3 slow stretches with the patient to determine the angle and torque through the whole elbow extension, followed by 3 fast stretches to detect the angle and torque at the catch, and then the therapist will determine the severity level of the particular patient based on MAS level. The post-assessment includes the visualisation of the torque and elbow angle side by side for analysis purpose. Figure 3 shows the instance of conducting the assessment on the patient by a trained clinician.

IOP Conf. Series: Materials Science and Engineering **824** (2020) 012013 doi:10.1088/1757-899X/824/1/012013

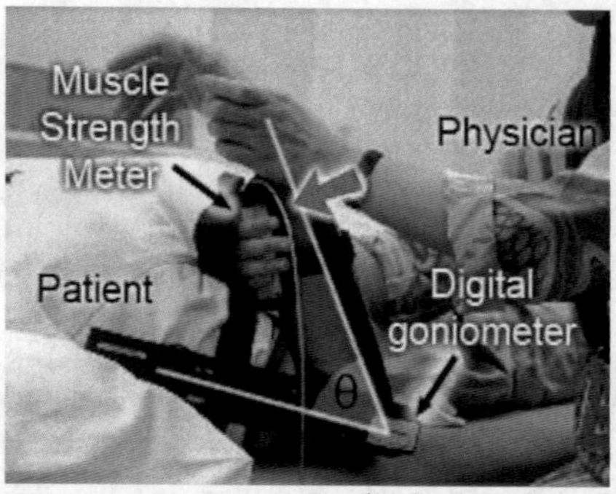

Figure 3. Measurement system attached on patient for assessment

3.2 Simulation program

The mathematical model produced by the HESS team [9] is reconstructed in MATLAB accordingly for simulation purpose. The mathematical model is mainly divided into three main phases according to their definition, which is the pre-catch phase, catch phase, and post-catch phase. The three main equations for the torque are connected as a piecewise function. The three main equations are as followed:

$$\tau_{pre} = m\ddot{\theta} + b\dot{\theta} + k\theta \quad (1)$$

$$\tau_{catch} = H\dot{\theta}_{catch_{init}}\delta(t) + \tau_{pre\_end} \quad (2)$$

$$\tau_{post} = m\ddot{\theta} + b\dot{\theta} + k_{post}(\theta - \theta_{pos\,t_{init}}) \quad (3)$$

The symbols above are as followed: τ_{pre} is pre-catch torque, τ_{catch} is catch torque, τ_{post} is post-catch torque, θ is the angle of elbow, m is mass of the forearm, b is damper constant, k is spring stiffness, k_{post} is spring stiffness in post-catch phase, H is the catch torque constant and $\delta(t)$ is the coefficient constant.

Some of the variables in the equations above are not clarified in this paper as all the declaration has been made by the authors in [9]. The three equations listed are only for reference purpose in this paper. All the constants in the equations for the simulation are according to the figures given by the original paper as well.

A MATLAB script is used to simulate the control system of the upper limb spasticity simulator. The main input of the script (angle) is set based on the clinical data obtained from previous study. Microsoft Excel is used for storing the input data and the program will read this data by extracting the specific Excel files.

Numerical differentiation is used to find a numerical value of a derivative of given function at a given point. In the ideal situation, the derivative of a certain mathematical function $f(x)$ can be estimated directly using the common differentiation method. However, there are always cases that a specific formula for $f(x)$ is not known in the real engineering applications. There are only a set of data points acquired to represent the dependent relationship between x and $f(x)$. Hence, the finite difference approximation can be used to figure out the estimate, $f'(x)$ when estimating the rate of change of $f(x)$ with respect to x. There are basically three ways to calculate the approximating derivatives, i.e. forward, backward and central difference method. For this study, 3-points backward difference method is used. The first derivative and second derivative of this method is written as follow.

$$f'(x) \approx \frac{3f(x) - 4f(x-h) + f(x-2h)}{2h} \tag{4}$$

$$f''(x) \approx \frac{f(x) - 2f(x-h) + f(x-2h)}{h^2} \tag{5}$$

where h = interval gap

Due to the fact that the collected clinical data is still insufficient for a thorough and complete simulation for all MAS level, only MAS level 1+ will be focused in this study. A total of 3 patients with ULS of MAS level 1+ have been engaged for data collection session, and the data are used for the simulation purpose in this study.

4. Result

The simulated results are compared with the real clinical data to assess and verify the applicability of the mathematical model in the context of Malaysian patients. 2 clinical data samples of MAS level 1+ are used for the comparison purpose. The main parameter that indicates spasticity is the catch angle. However, the definition of catch according to the professional clinician is not something that is obvious in the data set due to its qualitative nature and it is more sensational rather than by certain parameters. Hence, the peak torque and the peak torque angle (angle at which the peak torque occurs) are identified for verification purpose instead of the catch angle.

Three main datasets are collected from the clinical data measurement: torque, elbow angle, and their corresponding time. For every patient, 3 slow stretches and 3 fast stretches are applied on each clinical data collection session, thus there are 3 sets of data collected from each patient for the fast extension. The average values of peak torque and angle of peak torque are recorded in Table 2.

Table 2. Comparison of clinical data and simulation results

Data	Average Peak Torque [Nm]			Average Angle of Peak Torque [°]		
	Real	Sim	Difference	Real	Sim	Difference
1	13.287	11.642	-1.645	134.20	107.37	-26.83
2	12.052	15.734	+3.682	169.04	120.18	-48.86

From Table 2, it is seen that the mathematical model is able to imitate the peak torque and angle of peak torque with certain tolerance. For Patient 1, the simulated peak torque is 1.645 Nm short of the real torque value and it occurs 26.83° earlier. As for Patient 2, the simulated peak torque is 3.682 Nm

IOP Conf. Series: Materials Science and Engineering **824** (2020) 012013 doi:10.1088/1757-899X/824/1/012013

higher, and it occurs 46.86° earlier than the real clinical data. The elbow angle corresponding to the torque value could be referred to in the subplot immediately below the torque figure.

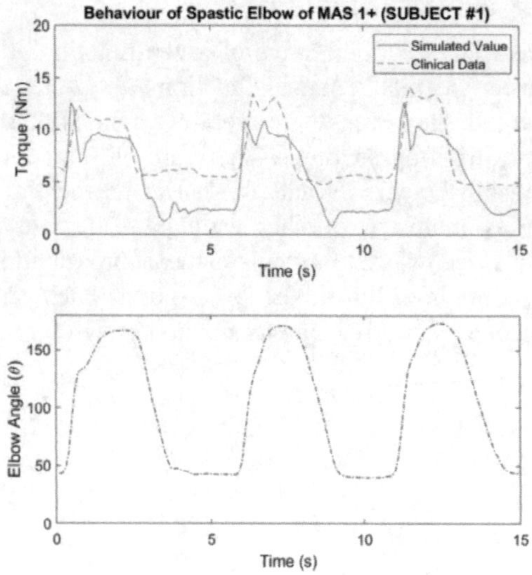

Figure 4. Graph of Torque and Elbow Angle against Time for Patient 1

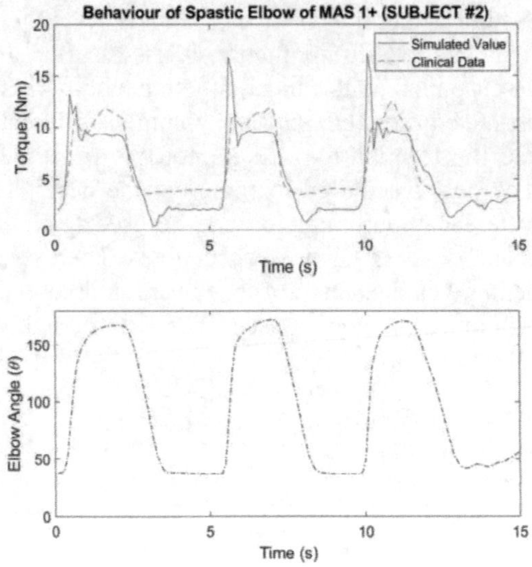

Figure 5. Graph of Torque and Elbow Angle against Time for Patient 2

From the graphs plotted, it is observed that the mathematical model is able to reproduce the trend of the slow stretches and fast stretches with the bell-shape curve for the slow stretches and the sudden surge for the fast stretches. It is observed that the simulated slow stretches (the first 3 torque hikes) and simulated fast stretches (the subsequent torque hikes) are able to follow the increment and decrement trend albeit the difference in magnitude. The shape of slow and fast stretches is able to be perceived clearly from the graphs.

Overall, the simulation results based on the clinical data collected from Malaysian patients are capable of imitating the real clinical data with some tolerance. The simulated results also resemble the definition according to [15] which stated that for MAS level 1+, the increase in muscle tone and catch would happen at the second half of the range of motion (ROM), and will be followed by minimal resistance throughout the remaining ROM.

5. Conclusion

In order for patients with ULS to return to normal activities of daily living, it is very important to have more trained professional therapists and clinicians to cater to the demands in physical and rehabilitation medicine, which in turn requires better medical education as the foundational support. Thus, ULS part-task trainer would be a very vital tool in producing healthcare providers of higher quality by allowing a safe zone for numerous practices for the trainees. In producing a ULS part-task trainer with higher fidelity, the overall program needs to be improved so that the trainees could undergo a top-tiered training before being launched into clinical training. In the current program employed by BITA (ULS part-task trainer previously developed), the value of torque is directly mapped onto the elbow angle as a discrete function and value, which is primitive in nature.

In this study, the mathematical model produced by other researchers is being verified by using the real clinical data collected from the Malaysian patients. The clinical data used are of MAS level 1+ as the clinical data of other MAS levels are insufficient. The mathematical model is being programmed into MATLAB script for simulation purpose with the angle of elbow as the input and the torque value produced as the output. From the simulated results, it can be concluded that the trend for slow and fast stretches is able to be reproduced successfully albeit the tolerance in the difference of value of peak torque and angle of elbow at which the peak torque occurs.

In the future study, more clinical data will be collected so that simulation for other MAS levels could be carried out for a more thorough simulation and verification purpose.

Acknowledgments
Authors wishing to acknowledge assistance by Ministry of Energy, Science, Technology, Environment and Climate Change [Ref. IF1118C1042], Universiti Tun Hussein Onn Malaysia [Ref. GPPS-U804], Universiti Teknologi MARA [Ref. 600-IRMI/DANA KCM 5/3/ LESTARI (133/2017) & 600-IRMI/PERDANA 5/3 BESTARI (087/2018)] and the German Federal Ministry of Education and Research under the Research Alliance for Intelligent Systems in Medical Technology in Malaysia (RAISE-MED). The clinical data collection is conducted under the Ethics Approval registered with National Medical Research Register (NMMR) Malaysia with Approval Number: NMRR-13-1384-18681.

References

[1] Jonghyun Kim, Hyung-Soon Park and Damiano D L 2011 Accuracy and reliability of haptic spasticity assessment using HESS (Haptic Elbow Spasticity Simulator) *2011 Annual International Conference of the IEEE Engineering in Medicine and Biology Society* (IEEE) pp 8527–30
[2] Cz N A, Komeda T and Low C Y 2012 Design of Upper Limb Patient Simulator *Procedia Eng.* **41** 1374–8
[3] Yee J, Low C Y, Hanapiah F A, Che Zakaria N A, Mohammad U and Abd Rahman R 2017 System-level design of a cloud-based training device for upper limb spasticity rehabilitation *2017 IEEE 3rd International Symposium in Robotics and Manufacturing Automation (ROMA)* (IEEE) pp 1–6
[4] Datta R, Upadhyay K and Jaideep C 2012 Simulation and its role in medical education *Med. J. Armed Forces India* **68** 167–72
[5] Jones F, Passos-neto C E and Braghiroli O F M 2015 Simulation in Medical Education : Brief history and methodology *Princ. Pract. Clin. Res.* **1** 56–63
[6] Takahashi Y, Komeda T, Koyama H, Yamamoto S-I, Arimatsu T, Kawakami Y, Inoue K and Ito Y 2011 Development of an upper limb patient simulator for physical therapy exercise *2011 IEEE International Conference on Rehabilitation Robotics* vol 2011 (IEEE) pp 1–4
[7] Grow D I, Mengnan Wu, Locastro M J, Arora S K, Bastian A J and Okamura A M 2008 Haptic Simulation of Elbow Joint Spasticity *2008 Symposium on Haptic Interfaces for Virtual Environment and Teleoperator Systems* (IEEE) pp 475–6
[8] Hyung-Soon Park, Jonghyun Kim and Damiano D L 2011 Haptic recreation of elbow spasticity *2011 IEEE International Conference on Rehabilitation Robotics* vol 221 (IEEE) pp 1–6
[9] Hyung-Soon Park, Jonghyun Kim and Damiano D L 2012 Development of a Haptic Elbow Spasticity Simulator (HESS) for Improving Accuracy and Reliability of Clinical Assessment of Spasticity *IEEE Trans. Neural Syst. Rehabil. Eng.* **20** 361–70
[10] Park J-H, Lee K-J, Yoon Y-S, Son E-J, Oh J-S, Kang S H, Kim H and Park H-S 2017 Development of elbow spasticity model for objective training of spasticity assessment of patients post stroke *2017 International Conference on Rehabilitation Robotics (ICORR)* (IEEE) pp 146–51
[11] Fujisawa T, Takagi M, Takahashi Y, Inoue K, Terada T, Kawakami Y and Komeda T 2007 Basic Research on the Upper Limb Patient Simulator *Proc. 2007 IEEE 10th Int. Conferece Rehabil. Robot.* **00** 48–51
[12] Takahashi Y, Yamamoto S-I, Inoue K, Komeda T, Arimatsu T, Ito Y, Koyama H and Kawakami Y 2012 Development of a patient simulator for physical therapy exercise of the upper limb *2012 9th France-Japan & 7th Europe-Asia Congress on Mechatronics*

IOP Conf. Series: Materials Science and Engineering **824** (2020) 012013 doi:10.1088/1757-899X/824/1/012013

(MECATRONICS) / 13th Int'l Workshop on Research and Education in Mechatronics (REM) vol 2011 (IEEE) pp 391–6

[13] Che Zakaria N A, Komeda T, Low C Y and Inoue K 2013 Emulation of muscle tone of upper limb spasticity and rigidity *2013 6th International IEEE/EMBS Conference on Neural Engineering (NER)* (IEEE) pp 1590–3

[14] Othman N A, Che Zakaria N A, Low C Y, Hanapiah F A, Komeda T and Inoue K 2015 Towards a Clinically Compliant Upper Limb Part-Task Trainer in Simulated Learning Program *J. Teknol.* **76** 71–6

[15] Othman N A, Idris F, Zakaria N A C, Hanapiah F A and Low C Y 2016 Supporting clinical evaluation of upper limb spasticity with quantitative data measurement in accordance to the Modified Ashworth Scale *2016 IEEE EMBS Conference on Biomedical Engineering and Sciences (IECBES)* (IEEE) pp 731–6

[16] Idris F, Zakaria N A C, Low C Y, Hanapiah F A and Othman N A 2017 System Integration of an Upper Limb Disorder Part-Task Trainer with PC-based Control *Procedia Comput. Sci.* **105** 328–32

[17] Goniometer (Order Code GNM-BTA)

[18] SensorDAQ Specifications

[19] Electrode Amplifier (Order Code EA-BTA)

IOP Conf. Series: Materials Science and Engineering **824** (2020) 012014 doi:10.1088/1757-899X/824/1/012014

Production of disposable takeaway food container from areca leaf sheath (ALS) as sustainable packaging material

Oi Ka Jong[1], Tze Mi Yong[1,2] and Muharam Rafidi Jaafar[1]

[1] Innovative Manufacturing Technology Research Group (IMT), Advanced Technology Centre (ATC), Department of Mechanical Engineering Technology, Faculty of Engineering Technology, Universiti Tun Hussein Onn Malaysia (UTHM), Pagoh Edu Hub, KM 1, Jalan Panchor, 84600 Pagoh, Muar, Johor, Malaysia

E-mail: tmyong@uthm.edu.my

Abstract. Areca leaf sheath has been used traditionally to produce disposable plates and bowls in India for decades. This project aims to produce disposable takeaway food containers by using the fallen Areca leaf sheath. In order to achieve the aim, a press mold is self- designed and fabricated, and installed into a disposable paper plate press machine to form the sheath into container shape. The density of the sheath is 0.423 g.cm^{-3} and 0.391 g.cm^{-3} before and after press respectively. The moisture content percentage obtained before heat press is 19.19%, while after press is 5.01%. The sheath at its sheath line direction shows higher stiffness, which is 416.9g.cm, compared to the perpendicular sheath direction of 65.1g.cm. The sheath has a water absorption rate of 0.021g.min^{-1}, which fulfils the purpose of disposable food packaging to contain foods with soup or sauce.

1. Introduction

Disposable tableware or dining ware is the articles used for eating a meal such as crockery, cutlery and others which are meant to be thrown away after use. Today, because of its single-use purpose, disposable tableware is low cost, convenient for takeaway, and easy to be disposed after used. For these reasons, disposable foodservice wares are preferred over the reusable ones by both the food servers and consumers[1-2]. Among the globally most used tableware are disposable cups, which have alone contributed over 37% of all foodservice disposables, with a total of 500 billion pieces to be sent to landfill yearly as reported by The Freedonia Group in 2010 [3]. Such massive quantity of trash alarms the environmentalists to study and even to raise awareness among the public regarding the disastrous impacts that the disposables can cause to our Mother Earth. The most direct solution is to stop the usage of the disposables, and be replaced by reusable dining ware. This solution is even initiatively employed by Starbucks, the leading global beverage brand in collaboration with researcher from Iowa State University [2].

In contrast to the encouragement in usage of reusable dining ware, there are studies suggesting that disposable dining ware might have more eco merit than reusable dining ware in terms of manufacturing energy and wash energy consumption. The reusable plates or cups need to be reused for a certain times to outweigh the impact that disposables bring to the environment, which is known as

[2] To whom any correspondence should be addressed.

the 'break-even point' [4-6]. Although being criticized widely, the initiator of such argument, Martin B. Hocking [4-5] emphasized that reusable dining ware is suitable for regular usage basis; but for one-time events like large parties where the return rate is low, disposable dining ware is still a better option. Since the disposable dining ware unavoidably still plays an important role in people's life, researches have been conducted in order to replace the conventional plastic disposables by using other materials such as paper, biodegradable plastics, bio-based plastics and even just leaves. In this project, the Areca leaf sheath (ALS) is proposed as a sustainable raw material to produce disposable takeaway food container.

ALS is the agro waste from the areca palm plantation, which the palm is scientifically known as Areca catechu L.. Due to its slow decomposition rate, it is usually gotten rid of by open burning [7-8]. The sheath of an adult is concaved in the center. The structure, composition and appearance of the sheath are heterogeneous. The outer surface of the ALS is greenish or brown, waxy and tough, while the inner surface is creamy in color and has a natural glossy finish [9]. ALS is generally a hard natural material, with tensile strength of 163.36 MPa along the grain direction, and 0.18 MPa along the transverse grain direction [10]. The fiber of ALS also contains higher amount of lignin (38.68%), lower amount of cellulose (26.40%) and hemicellulose (16.50%) compared to other commercialized natural non-wood fibers like jute, hemp, flax and *etc* [11-12]. The high lignin content gives ALS high stiffness and rigidity [13].

Due to the rigidity and high mechanical strength of ALS, it has been used as a biodegradable raw material to produce disposable plates and bowls for decades in India through compression molding process [7-8,10]. After washed, the ALS is subjected into the hydraulic press machine at which the upper die and lower die temperature is set to 110-115 °C and 35-45 °C respectively, and stamped for 40-60 seconds [8]. It should be noted that the moisture content of the sheaths should be maintained at above 5% to prevent cracking during the stamping process. However, if the moisture content exceeds 30%, the product will be difficult to remain dimensionally stable, and fungi be most likely to grow during the storage time of the products. Besides, the products are also sanitized when they are subjected to high temperature. It is also recommended that the moisture content of the raw sheath should be maintained at 14-17% before production [8,10]. With these production parameters as guidelines, this project attempted to produce a self-designed disposable takeaway food container with burger box size. The design of the container aims to explore a more complicated structure for ALS disposable products compared to the existing ones. Some physical and mechanical properties of ALS will also be investigated in this study.

2. Methods

2.1 Collection of ALS

The fallen ALS were collected from the areas of Batu Pahat and Muar, Johor in Malaysia. These areca palm trees are used as decoration plants. The collected ALS were then washed with running water and dried under the sun, before being stored in an air-conditioned laboratory.

2.2 Density of ALS

Density test was done in accordance to standard of TAPPI T 410 om-08. This standard procedure of paper grammage test was adopted for measurement of ALS density due to its flat, thin and hydrophilic characteristics which are similar to that of a paperboard. The density of the ALS was measured before and after the compression molding production process of food container. 10 replicates of 8 cm ×8 cm were measured.

2.3 Moisture content of ALS

The initial moisture content of 10 ALS samples (8 cm × 8 cm) were measured as well as the moisture content after the compression molding process. The procedure was in accordance to standard of TAPPI T 550 om-08 which uses the oven drying method.

IOP Conf. Series: Materials Science and Engineering **824** (2020) 012014 doi:10.1088/1757-899X/824/1/012014

2.4 Taber bending stiffness of ALS

The test was carried out by referring to TAPPI T 489, on the ALS to find out the bending moment required to deflect it. By using a Taber Stiffness Tester (Model 150-E, Taber), 10 specimens for both parallel and perpendicular grain direction of ALS were tested at Range 4 of Taber Stiffness unit. Then, the bending moment and resistance to bending are calculated by using equation (1) and (2):

$$\text{Bending moment (mN.m)} = \text{Average instrumental readings in stiffness units (g.m)} \times 0.098066 \quad (1)$$

$$\text{Resistance of bending (mN)} = \frac{\text{Bending moment (mN.m)}}{\text{Length (m)}} \quad (2)$$

2.5 Design of ALS food container

The design of the food container was done using the CAD software SolidWorks 2016. The shape of the takeaway container is designed to be having only a single cavity (figure 1) instead of the clamshell shape with 2 joined cavities like the conventional takeaway containers. To make a complete container, two cavities are needed to combine with each other by the tab-and-slot mechanism. A post-process to cut a gap on the slot of the container is needed with this design. This food container design is intended to save the raw material of press mold.

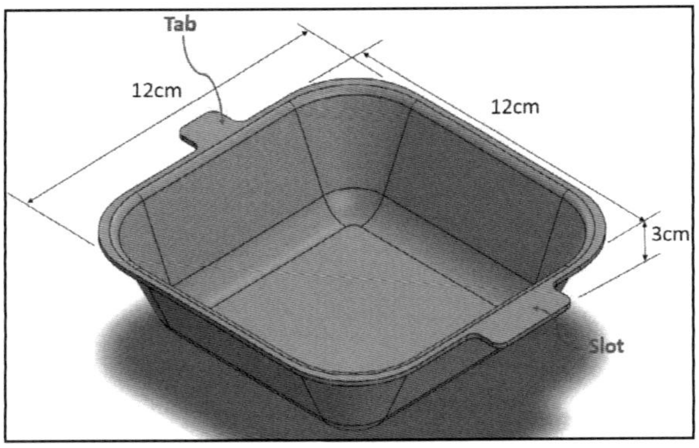

Figure 1. Design of ALS food container.

2.6 Design and fabrication of press mold

The press mold was designed using SolidWorks 2016, with the 'Core and Cavity' feature. Then, the designed mold is fabricated by using a three-axis CNC milling machine (Hwacheon Vesta 660) after the fabrication process was simulated with MasterCAM 2017 software. The material chosen to make the press mold is aluminum due to its more affordable cost, higher ductility, and light weight for a prototype making purpose. End milling processes were performed by the CNC milling machine to produce both punch and die of the press mold. The fabricated mold is as shown in figure 2. The mold was then installed on an existing press machine which was used to produce disposable paper plates.

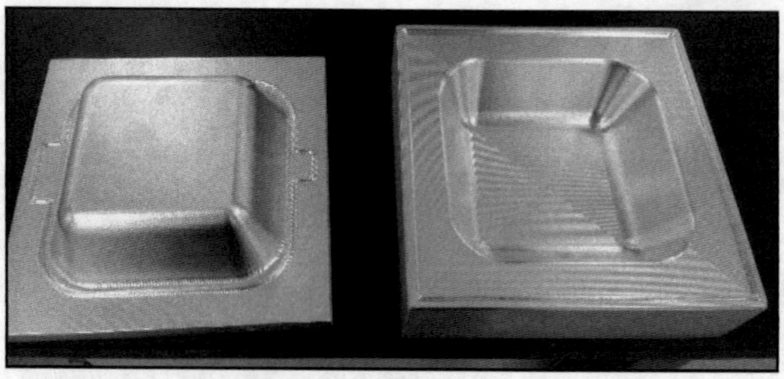

Figure 2. The fabricated punch (left) and die cavity (right).

2.7 Production of ALS food container

A few trial runs with the newly fabricated mold were carried out by referring to the parameters recommended by Kalita et al. [10]. Finally, the temperature of the punch was set to be 100±10°C, whereas the cavity was 40±10°C. The pressing time duration was controlled to be 2 minutes. After the shape of the product is formed, post-processes like trimming the excess part of the containers and cutting the gap on the slot were done. Two containers were combined to form a complete product.

2.8 Water absorption rate of ALS containers

The ALS disposable containers are expected to be able to contain foods with gravy, thus it should be able to withstand water penetration. 4 containers were filled with 50 ml of water, and the water would be removed for every 30 minutes to weigh the wetted containers. This weighing process repeated till 3 hours and 30 minutes. Then, the water absorption rate of the ALS containers was calculated using equation (3) below:

$$\text{Water absorption rate (g.min}^{-1}) = \frac{\text{Total weight increased (g)}}{\text{Time (min)}} \tag{3}$$

3. Results and discussion

3.1 Density

The average thickness of 10 ALS samples before being pressed is 1.08 mm, while the thickness after press is 1.00 mm. Next, the average weight ALS specimens before press is 2.93g, and 2.50g after press. Hence, the density is 0.423 g.cm^{-3} and 0.391 g.cm^{-3} before and after press respectively. Although the density of ALS is higher than that of polystyrene (0.045g.cm^{-3}), which may increase its transportation cost, the obvious advantage of being eco-friendly is able to overcome this shortcoming of ALS.

3.2 Moisture content

The range of the moisture content of the ALS is between 16.46 to 22.02%, with an average value of 19.19%. These values are similar to those that are reported by Kalita et al. [10], that is 14-17%. After the sheath is pressed, the average moisture content is 5.01%.

3.3 Bending stiffness

As shown in table 1 below, ALS required higher moment to be bent at the parallel grain direction (40.88 mN.m) compared to the perpendicular grain direction (6.38 mN.m). In order to bend the parallel grained ALS, 817.60 mN is needed while 127.60 mN is need to bend the perpendicular grained ALS.

Table 1. Bending Moment and Resistance to Bending of ALS

Properties	Parallel	Perpendicular
Stiffness unit, g.cm	416.90	65.10
Bending moment, mN.m	40.88	6.38
Resistance to bending, mN	817.60	127.60

3.4 Production of ALS food container

A complete ALS food container was produced (figure 3) by compression molding process. However, as the press machine was used to produce disposable paper plate, the machine does not have enough pressure to fully compress the sheaths. Thus, as illustrated in figure 4, the ALS that was folding up at the edge of the container was not completely flattened by the compressive force of the press machine. In spite of the imperfect finishing at the edge of the containers, the produced ALS containers are able to fulfill its main function to contain and enclose foods.

Figure 3. A complete ALS container.

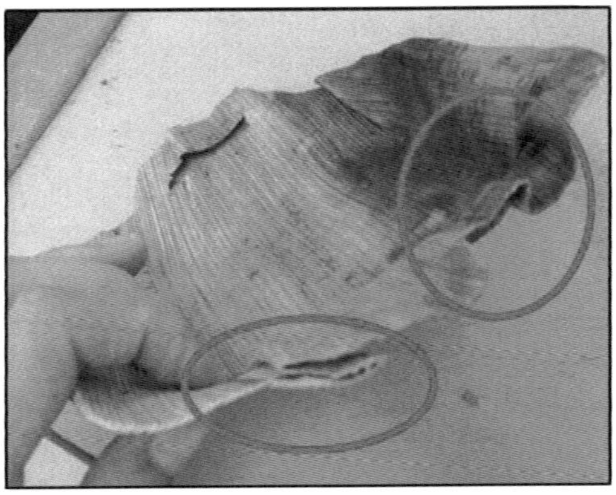

Figure 4. The incomplete folds of ALS container.

3.5 Water absorption rate

Throughout the test, the average weight gained by the specimens is 3.85g in 3 hours, which gives a water absorption rate of 0.021 $g.min^{-1}$ (figure 5). It is also observed that when the water remains in the container for the next 3 hours, the water still did not penetrate through any of the specimen. The sheath container is able to fulfill the requirement of a disposable packaging with its low water absorption rate to withstand soup or sauce penetration through the material.

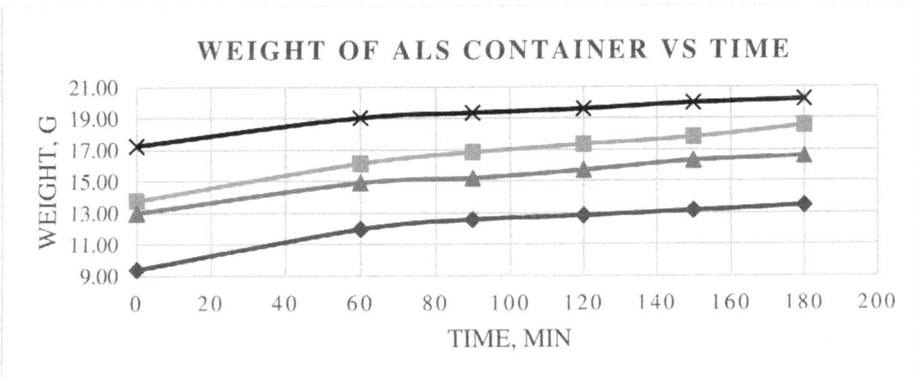

Figure 5. Graph of weight of ALS container (g) against time (min).

4. Conclusion

The ALS takeaway disposable food containers are successfully produced from the fabricated mold. These containers are able to fulfill the purposes of disposable food packaging. The pressing temperature of the punch is set to be 100±10˚C, whereas the cavity is 40±10˚C. The pressing time duration is 2 minutes. The density of the raw sheath is 0.424 g.cm^{-3}, after pressed it is 0.391 g.cm^{-3}. The average moisture content of the sheath before press is 19.19%, while after press is 5.01%. Besides, the sheath line direction of the areca leaf sheath has greater stiffness, which is 416.9g.cm compared to the transverse sheath line direction, that is 65.1 g.cm. The areca leaf sheath container is able to withstand water penetration for more than six hours, having a water absorption rate of 0.021 g.min^{-1}, which makes it a suitable disposable food packaging material to contain foods with soup or sauce. For future research, the moisture content of the ALS can be controlled to reduce cracking and folding of products. Since a simple paper plate press machine was used for this project, a hydraulic press can be used in the future work for a greater compressive force.

5. Acknowledgement

First and foremost, the authors want to thank Universiti Tun Hussein Onn Malaysia (UTHM), Malaysia in providing the fund and necessary facilities needed to complete this project.

6. References

[1] Fang W T, Ng E, Wang C M and Hsu M L 2017 Normative Beliefs, Attitudes, and Social Norms: People Reduce Waste as an Index of Social Relationships When SPending Leisure Time *Sustainability* **9** 1696

[2] Lee J 2015 A study for increasing reusable cup consumption in the coffee industry : focused on behavior change with motivation *Graduate Theses and Dissertations* (Iowa: Iowa State University Digital Repository) 14847

[3] Woods L and Bakshi B R 2014 Reusable vs. disposable cups revisitede: guidance in life cycle comparison addressing scenario, model, and parameter uncertainties for the US consumer *Int. J. Life Cycle Assess.* **19** 931–940

[4] Hocking M B 1994 Disposable cups have eco merit *Nature*, **369** 107

[5] Sheehan B, Gordon M and Sommer S 2017 *Greenhouse Gas Impacts of Disposable vs Reusable Foodservice Products: Literature Review and Inventory* (Clean Water Fund)

[6] Hocking M B 1994 Reusable and Disposable Cups : An Energy-Based Evaluation *Environ. Manage.* **18** 889–99

[7] Bavappa K V A, Nair M K and Kumar T P 1982 *The Arecanut Palm* (Kerala, India: Central Plantation Crops Research Institute)

[8] Kalita P, Dixit U S, Mahanta P and Saha U K 2008 A novel energy efficient machine for plate manufacturing from areca palm leaf sheath *J. Sci. Ind. Res. (India).* **67** 807–11

[9] Nair K P P 2010 *The Agronomy and Economy of Important Tree Crops of the Developing World*, 1st ed (New Delhi, India: Elsevier Inc.)

[10] Kalita P, Dixit U S, Mahanta P and Saha U K 2006 Effect of Moisture and Temperature on Arecanut Leaf Sheath Products *Proc of the 3rd BSME-ASME Int. Conf. on Thermal Engineering*

[11] Sadrmanesh V and Chen Y 2018 Simulation of tensile behavior of plant fibers using the Discrete Element Method (DEM)," *Compos. Part A Appl. Sci. Manuf.* **114** 196–203

[12] Nagaraja R, Gurumurthy B R and Shivanna M B 2014 Bio Softening of Arecanut Waste Areca Husk, Leaf and Leaf Sheath For Value Added Compost *IMPACT Int. J. Res. Appl. Nat. Soc. Sci.* **2** 105–12

[13] Chen H 2014 Chemical Composition and Structure of Natural Lignocellulose *Biotechnology of Lignocellulose* (Dordrecht: Springer) pp 25–71

IOP Conf. Series: Materials Science and Engineering **824** (2020) 012015 doi:10.1088/1757-899X/824/1/012015

Physical Properties and Hardness of Treated Amorphous Silica as Reinforcement of AA7075 Recycled Aluminum Chip

Nurul Farahin Mohd Joharudin[1,2], Noradila Abdul Latif[1,2,*], Mohammad Sukri Mustapa[1,2], Nur Azam Badarulzaman[2], Muhammad Faisal Mahmod[2]
[1]Structure Integrity Monitoring Research Group (SIMReG),
[2]Faculty of Mechanical and Manufacturing Engineering,
Universiti Tun Hussein Onn Malaysia, 86400 Parit Raja, Johor.

E-mail: noradila@uthm.edu.my

Abstract. Solid state recycling techniques of waste aluminum materials without melting the material is a process that consumed less energy consumption. The reuses of waste materials such as aluminum scrap and rice husk ash lead to the friendly environmentally and low cost of production. Physical properties and hardness of treated amorphous silica as reinforcement of AA7075 recycled aluminum chip were investigated. Recycled aluminum sample AA7075 reinforced treated amorphous silica by rice husk ash was prepared by the technique of powder metallurgy (PM) owing of reduced energy consumption and operating cost in comparison with conventional cast-recycling. The presence of silica exists in rice husk ash obtained after through chemical treatment and burning process of rice husk at temperatures of 700°C for 2 hours. The effect of aluminum recycled chip with rice husk ash composite with composition of 2.5 wt.%, 5 wt.%, 7.5 wt.%, 10 wt.% and 12.5 wt.%, on physical properties and hardness were investigated. Based on investigation to AA7075 recycled aluminum chip reinforced treated amorphous silica by rice husk ash, the porosity and water absorption of composite of metal matrix is increase with increasing of composition of treated rice husk ash as the density shows decrease when the treated rice husk ash is added up to 5 wt.%. The addition of treated rice husk ash has improve the hardness of composite of metal matrix from 53.49 Hv of 100 wt.% of AA7075 recycled aluminum chip to 69.53 Hv of composite of metal matrix at 10 wt.% of composition of treated rice husk ash. Based on this research, treated amorphous silica as reinforcement has enhance the physical properties and hardness of recycled aluminum chip sample in composition up to 10 wt.% of treated amorphous silica.

1. Introduction

Aluminum metal is a very light material with a specific weight of 2.7 gm/cm^3, heavily nonferrous metals used in the world and it is about third most abundant element in the earth's crust [1,2]. Waste material has becomes major issue as it's produced in manufacturing industry as aluminum scrap. Aluminum scrap obtained as secondary aluminum promise environmentally friendly and low cost of production than the cost production by conventional methods [3-5].

During the production of products, after the machining process, different sizes and types of chips are generated [6]. Conventional recycling leads oxidation loss of metals and also increases labor and energy costs, only 5 percent of the energy needed for ore aluminium production to be recycled [3]. Consequently secondary aluminum in many countries has been accepted.

IOP Conf. Series: Materials Science and Engineering **824** (2020) 012015 doi:10.1088/1757-899X/824/1/012015

Rice husk and its ash are used as a potential reinforcement additive to replace conventional forms of silica as a major source [7]. Crystalline silica was obtained at burning temperature of 1100°C and 1200°C, while silica was observed at burning temperatures of 900°C and 1000°C in transitional phase amorphous to crystalline and silica were remains in amorphous phase at burning temperatures of 700°C and 800°C [8-10].

Based on the study of AA7075, the composite material of the composite of metal matric can be improved with an addition amorphous silica with appropriate composition treated amorphous silica to composite of metal matrix. Thus, silica produced by rice husk is excellent natural source that can be able to replace conventional silica sources, where agricultural waste can be change into beneficial agricultural raw material.

2. Experimental Material and Procedure

2.1. Aluminum Matrix Material

The matrix material used in this study is aluminum AA7075 alloy. The theoretical density of material is 2.81 g/cm^3 where the element composition of aluminum AA7075 alloy (wt.%) is shown in Table 1.

Table 1. Element composition of aluminum AA7075 alloy (wt.%) [10]

Al	Zn	Cu	Si	Cr	Others
87.18	9.49	2.59	0.31	0.28	0.15

2.2. Reinforced Material

Rice husk obtained from industry of rice husk company, Nano Siltech Sdn Bhd, was used as reinforcing agent. The rice husk was burned at 700°C using chemical treatment (hydrochloric acid) to fasten the burning process (amorphous phase). The chemical element composition (wt.%) of treated amorphous silica obtained by x-ray fluorescence analysis (XRF test) as shown in Table 2.

Table 2. Chemical element composition of treated amorphous silica (wt %)

SiO_2	C	K_2O	P_2O_5	CaO	MgO	SO_3	Cl	Fe_2O_3	Al_2O_3
87.52	0.11	4.83	3.52	1.44	1.28	0.54	0.13	0.39	0.24

2.3. Chips AA7075Preparation

The high-speed CNC milling machine Mazak Nexus 410A-II CNC Mill produced the aluminum chip at 1100 mm/min feed rate, 1 mm depth cut and 345.4 m/min cutting velocity. FRITSCH – ultrasonic cleaner Labarette 17 was then was used to clean the milled aluminum. The cleaning time used for the treatment of grease, oil and impurities was 1 hour with an acetone solution (CH_3COCH_3). Finally, it was used to dry the residual acetone from the chip by using a drying oven in 1 hour at 75ºC [11-14].

2.4. Rice Husk Ash Preparation

In preparing rice husk ash, the ball mill was used to generate smaller particle size by using porcelain jar for 10 minutes under wet milling cycles. FRITSCH analysette 3 vibrator apparatus was used to perform sieving process with duration of 30 minutes for separating the particle size of 63μm.

2.5. Mixing and Compaction Process

A machine of uniaxial hydraulic press in Carver model 3851-0, was used to compact the samples. The preparation of mixing samples as shown in Table 3. The aluminum chips AA7075 were then mixed with treated amorphous silica, and then shaped to the mold. The sample is in cylindrical shape with size of 10 mm height and 14 mm diameter. The load compaction and compaction duration were set at 9 tonnes and 20 minutes, respectively [15,16]. The die surface was cleaned with lubricants for easy removal of the samples, by spraying the surface.

Table 3. Composition of treated amorphous silica and aluminum AA7075 chip

Composites	Composition
A	100% AA7075 chip
B	97.5% AA7075 chip + 2.5% treated amorphous of silica
C	95.0% AA7075 chip + 5% treated amorphous of silica
D	92.5% AA7075 chip + 7.5% treated amorphous of silica
E	90.0% AA7075 chip + 10% treated amorphous of silica
F	87.5% AA7075 chip + 12.5% treated amorphous of silica

2.6. Sintering Process
Tube furnace is used for sintering process under inert gas (Argon gas) control. The heating rate, temperature and time were controlled for reproducible results. The sintering profile shown in Figure 1 was the temperature used to performed this process [10,15]. The zinc stearate was burned out at 300°C after 30 minutes and the sintering procedure began within 60 minutes at 552°C.

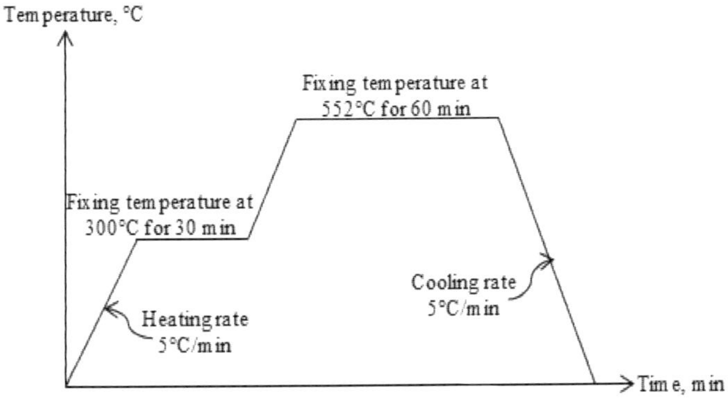

Figure 1. The sintering profile

2.7. Physical Test and Hardness Test
The physical test consists of density, porosity and water absorption of composite of metal matrixs is measured with Archimedes Principle which is using an electronic balance, Mettler Toledo German. The analysis testing was conducted followed by standard ASTM B328 for density and ASTM B962-17 for porosity.

Hardness test was conducted using Vickers method followed by standard ASTM E-384. Vikers hardness test machine equipped with a diamond indenter to measure the hardness. The measurements were performed with eight indenters to take the average.

3. Results and Discussion

3.1. Physical Properties
Archimedes method has been used to identify the relationship of physical properties between the density, apparent porosity and water absorption of composite of metal matrix at different composition of treated amorphous silica. Table 4 shows the results of density, porosity and water absorption for composite of metal matrix and Figure 2 shows the graph of density, porosity and water absorption for composite of metal matrix sample.

Table 4. Physical test.

Composites	Density (g/cm^3)	Porosity (%)	Water absorption (%)
A	2.34	13.4215	6.4071
B	2.47	9.8659	4.4146
C	2.49	9.3247	4.1098
D	2.44	15.3990	7.4164
E	2.42	14.5194	6.9874
F	2.38	15.4281	7.6196

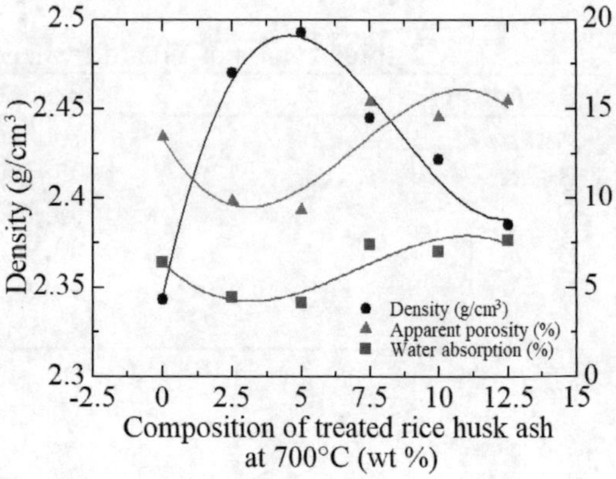

Figure 2. Graph of physical test.

Figure 2 observed that the density of composite of metal matrix were increase steadily by up to 5 wt.% at the composition of treated amorphous silica. The density of composite of metal matrix were decreased after being reinforced by more than 5 wt % of treated amorphous silica. It is certain that the density of composite of metal matrix was significantly dependent on the composition of treated amorphous silica. Meanwhile, the apparent porosity and water absorption were increase to the increasing in composition of treated amorphous silica. It can be concluded that the higher the addition composition of treated amorphous silica, the higher the pore formed.

3.2. Microhardness

The hardness of material was performed on the material surface using the Vickers hardness machine at small load of 1 N, as it is consists of a hard and precise head for easing the penetration into the sample material. The average of hardness was taken with eight times of indentation into each sample.

Figure 3 shows the relationship between hardness and different compositions of treated amorphous silica. It has been found that the hardness increases with an increasing mass fraction of the treated amorphous silica content, improving the hardness of the composite metal matrix. The hardness of composite of metal matrix has improved from 53.49 Hv of fully AA7075 recycled aluminum chip to 69.53 Hv of composite of metal matrix. The hardness reaches 69.52 Hv at 10 wt % of treated amorphous silica and it is higher than hardness of untreated amorphous silica that reaches 65.93 Hv at 10 wt % [10]. However, Saravanan et al., and Tiwari et al., reported that the hardness of matrix material reduced with silica composition due to poor bonding of the matrix and poor wettability of reinforcing rice husk material with aluminum material when the addition of silica were more than 12 wt.% [7,17].

Table 5. Microhardness data

Composites	Microhardness (Hv)	Std. dev.
A	53.4905	3.43413
B	55.8036	2.77232
C	63.2777	4.25264
D	64.5604	2.47375
E	69.5261	4.41494
F	65.7145	2.22012

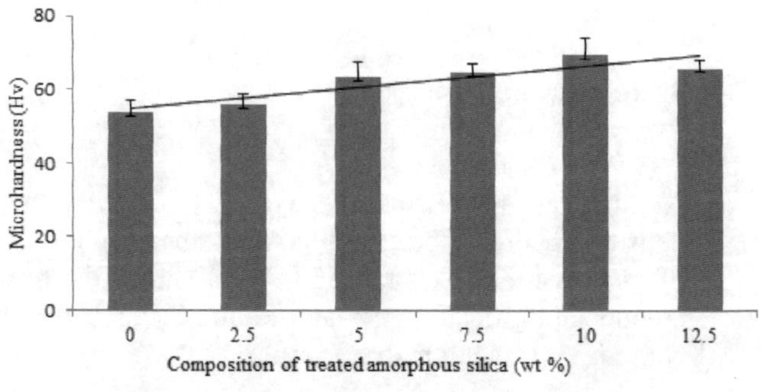

Figure 3. Microhardness (Hv) at different composition of treated amorphous silica (wt.%)

4. Summary

In conclusion, the summarized of current study is as follows:

1. Rice husk, industrial residue from paddy milling, which are containing high amount of silica can be used as aluminum matrix composites reinforcing agent.
2. Industrial wastes become economic resources if rice husks are used to produce composites. This also solves the issue of rice husk problem of storage and disposal waste.
3. The density of composite metal matrix has been risen to 5 wt% of treated amorphous silica and then reduced by enhanced mass fraction of treated amorphous silica. The porosity of composite metal matrix increase at increasing composition of treated amorphous silica.
4. The hardness of composite of metal matrixs was seen to be decreased with composite more than 12.5 wt% of treated amorphous silica by rice husk ash.

5. References

[1] P.G. Sheasby and R. Pinner 2001 The Surface Treatment and Finishing of Aluminum and Its Alloys, 6th Edition Finishing Publications Ltd pp.1-4
[2] J. Jenix Rino, D. Chandramohan, K.S. Sucitharan 2012 International Journal of Science and Research 1(3) 196-203
[3] J.Z. Gronostajski, H. Marciniak, A. Matuszak, M. Samuel 2001 Journal of Materials Processing Technology 119(1-3) 251-256
[4] J.B. Fogagnolo, E.M. Ruiz-Navas, M.A. Simon, M.A. Martinez 2003 Journal of Materials Processing Technology 143-144 792-795
[5] Z. Sherafat, M.H. Paydar, R. Ebrahimi 2009 Journal of Alloys and Compound 487 395-399
[6] Puga, H., Barbosa, J., Soares, D., Silva, F., and Riberio, S., 2009 Journal of Materials Processing Technology 209 5195-5203
[7] Srikant Tiwari, M.K. Pradhan 2017 Materialstoday: Proceedings 4(2) 486-495
[8] A.A. Boateng and D.A.Skeete 1990 Cement and Concrete Research 795-802
[9] N.A.A. Ismail, M. Azham Azmi, S. Ahmad, H. Taib 2014 International Conference on X-Rays & Related Technique in Research & Industry 2014 183-184
[10] N.F. Mohd Joharudin, N. Abdul Latif, M.S. Mustapa, M.N. Mansor, W.A. Siswanto, J. Murugesan, F. Yusof 2019 Materialwiss. Werkstoftech. 50 1-6
[11] M.I.A.Kadir, M.S.Mustapa, N.A.Latif, A.S.Mahdi, Procedia Engineering, 2017,184,687
[12] A.S. Mahdi, M.S. Mustapa, N.A. Latif, M.I.A. Kadir, M.A. Samsi 2017 International Journal of Engineering and Technology 8(6) 2582
[13] M.S. Mustapa, A.S. Mahdi, M.A. Lajis, M.W.A. Rashid 2016 International Journal of Mechanical and Industrial Technology 3(2) 33-40
[14] M.S. Mustapa, N.A. Latif, N.F.M. Joharudin, S. Mahzan, M.I. Masirin, N.A. Hamid 2018 Materials Science Forum 934 124
[15] A.S. Mahdi, M.S. Mustapa, M.A. Lajis, M.W.A. Rashid 2015 International Journal of Mechanical Engineering and Technology 6(9) 133-142
[16] A.S. Mahdi, M.S. Mustapa, M.A. Lajis, M.W.A. Rashid 2015 International Journal of Science and Research 4(9) 1759-1764
[17] S.D. Saravanan, M. Senthil Kumar 2013 Procedia Engineering 64 1505-1513

Acknowledgements

The authors acknowledged the support of the Universiti Tun Hussein Onn Malaysia for TIER 1 research grant No. U869, FRGS research grant No. K067 and the Ministry of Higher Education Malaysia (FRGS/1/2018/TK03/UTHM/03/17).

IOP Conf. Series: Materials Science and Engineering **824** (2020) 012016 doi:10.1088/1757-899X/824/1/012016

Optimization of Acoustical Properties Polyurethane (PU) Wood Fibres Foam Composites

Hanani Abdul Wahab[1], Anika Zafiah M. Rus[1], Ahmad Tajuddin Moen [1], Khairul Nizam Ngadimon[1], Fazimah Mat Noor[1]

[1]Sustainable Polymer Engineering, Advanced Manufacturing and Materials Center (SPEN-AMMC), Faculty of Mechanical and Manufacturing, Engineering Universiti Tun Hussein Onn Malaysia, 86400 Parit Raja, Batu Pahat, Johor, Malaysia

E-mail: nani@uthm.edu.my

Abstract. In this study, an optimization approach was proposed to improve acoustical behaviours of PU foams. The important parameters of PU foams: content of the combination of renewable monomer and epoxy as polyol, methylene diphenyl diisocyanate (MDI) as crosslinker. Their effects on sound absorption coefficient of PU foams were studied by using Taguchi methods. In addition, wood fibres were incorporated into PU foams as fillers to improve the acoustical properties of PU foams. Three controlled factors: the monomers, wood fibres and sample thickness with three levels for each factor were chosen and Taguchi method based on orthogonal array L9 (3^3) was employed to conduct the experiments. Based on the results of Taguchi's orthogonal array L9 (3^3), signal-to noise (S/N) analysis was used and developed to determine an optimal formulation of PU wood fibres foam composites. The results showed that the noise reduction coefficient (NRC) had showed an increased value from 0.4850 to 0.5550 positively improved the acoustic property and the best combination values of variables were monomer mixture of 25 gram, filler ratio of 3% wt and sample thickness of 30mm.

1. Introduction

Recently, sound absorbing materials constitutes one of the major requirements for human comfort and economical nowadays which required in modern technologies environment. The improvement of controlling noise by sound absorption contributes a great opportunity for acoustic attenuation method of various porous materials. Sound absorbing materials are classified on the ability of the material to absorb as much as sound wave and reflect as minimal as it could and at the same time transmit more of the waves [1]. Porous materials are excellent for sound absorption and good heat insulator. Its open pores permit confined airflow via the material thus absorbing sound and additionally preventing efficient heat exchange. Rus et al., [2] studied the porous materials with uniform and non-uniform pore structure gives different result in sound absorption coefficient at different frequency level. Flexible polyurethane foams (PU foams) with high porosity have been extensively used as noise control materials in many applications in industry due to effective sound absorption, light weight, excellent viscoelasticity and acoustic absorption properties [3,4].

[1] nani@uthm.edu.my

The acoustic panels that made from natural fibres are less hazardous to human health and eco-friendly than those made from common synthetic fibres. Hence, a rising concern for human health and safety issues has urged manufacturers and engineers to inquire alternative materials from natural fibres as a replacement for synthetic fibres. Natural fibers are essentially completely biodegradable and modern technical developments have made natural fiber processing more economical and environmentally friendly.

Many researchers [5-9] have examined and researched the suitability, competitiveness, and capabilities of natural fibres embedded in polymeric matrices. The researchers concentrated on the effect of the fibre surface modifications as well as manufacturing processes in improving fibre or polymer compatibility. On the other hand, some researchers studied and compared between different natural fibre composites and their stability in various applications [10]. Therefore, the effective sound absorption of any natural fibre polymer composite material can be achieved when it has a more tortuous path, higher surface area, higher flow resistivity, and low porosity within it at the optimal range.

Since, the issue of environmental and sustainability with climate change has become a major consideration in the development of new materials in the world, the wood fiber waste is the best filler to be used. Wood fiber is structural materials in plants and the most abundant biomass in earth. The use of wood fibers as fillers in polymeric composites show advantages as they are inexpensive, derive from renewable resources, present lower density than mineral fibers, undergo little damage during process and their disposal causes minor ecological impact [11].

Design of experiments (DoE) is a systematic approach to engineering problem-solving that applies principles and techniques at the data collection stage. Study by P. Muray et al., [12] highlighted the benefits of using DoE optimization during the development of new synthetic methodology. In other words, DoE method is used to find the cause and effect relationships which is significant to manage inputs and optimize the output prior going into production [13,14]. Meanwhile, Taguchi method is a broadly accepted method of DoE which has proven in producing high-quality products at subsequently low cost [15]. It provides a simple, efficient and systematic approach to estimate the effects of several variables separately, simultaneously or as combinations [16, 17]. Therefore, in this study, Taguchi method based on orthogonal arrays (OA) is a structured with an organized method has been applied to perform the present experiments and the sound absorption of PU foam composites was studied.

2. Methodology

2.1. Materials

Renewable bio-epoxy from the waste cooking oil were prepare at E1 Sustainable Polymer Engineering (AMMC), UTHM, Malaysia which comprises the acid-catalyst ring opening of the epoxides to form polyols [18]. Flexible polyol (Aura 2104 H3) and flexible MDI (diphenylmethane 4,4-diisocyanate) were purchased from Sykt Saintifik Bersatu (M) Sdn Bhd. Table 1 shows the materials used for preparation of PU foams. The wood dusts were obtained from Tukang Kayu A.Hamid Sdn Bhd, furniture industry from Batu Pahat, Johor With the size of wood fibers, 900-1200 μm, 3 different percentages has been selected which is 1 wt%, 3 wt% and 5 wt%. Figure 1 shows the photographs of (a) wood fibres and (b) the size of wood fibres under microscopy.

Table 1. PU foams formulation.

Materials	Content of each materials (gram)
Flexible polyol	25, 15, 5
Renewable Bio-epoxy	5, 15, 25
MDI	15
Wood Fibres	0.45, 1.35, 2.25

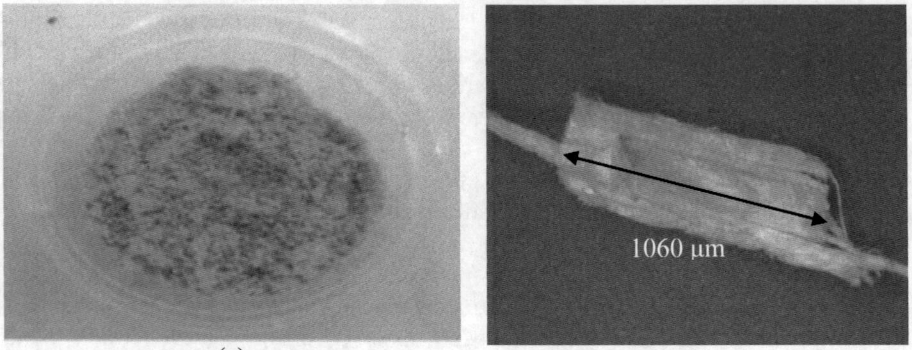

(a) (b)

Figure 1. Wood fibers prepared for PU foam composites.

2.2. Synthesis of PU foam composites

Firstly, flexible polyols were mixed with renewable bio-epoxy were prepared in the plastic container. Subsequently, wood fibres with (1%, 3% and 5%) of matrix weights ratios were added to the mixture and mixed for 5 minutes, afterward, croslinker, MDI was added to the wood fibres/polyols mixture and vigorous stirred for 15 second. Finally, the mixture was left at room temperature at room temperature to expand and cure for 24 hours. Upon curing, the samples were removed from the plastic container and were cut using knife and cylindrical shape of Teflon mold. Cylindrical samples with a diameter of 100 mm were used for the low frequency range (100–2000 Hz) and samples with 28 mm diameter were used for the high frequency range (200–6000 Hz). All the samples had a thickness of 10mm, 20mm and 30mm.

2.3. Experimental Design

The effects of process parameters in the formulation of PU foam composites and the optimum combination for acoustic properties of PU foam composites were studied by using Taguchi approach. The experiments were conducted with three controllable 3-levels factors and one objective. Table 2 shows three controlled factors, i.e., the polyol and renewable bio-epoxy mixture (i.e., A), the content of wood fibers (i.e., B) and the sample thickness (i.e., C) with three levels for each factor. The orthogonal arrays L9 (33) was chosen and then 9 experimental runs were carried out based on the design matrix shown in Table 3.

In engineering, noise reduction coefficient (NRC) is commonly used as an evaluation parameter to characterize sound absorbing materials. NRC is defined as arithmetic mean of the absorption coefficient at frequencies of 250, 500, 1000 and 2000 Hz, which is defined in the following equation,

$$NRC = \frac{\alpha_{250} + \alpha_{500} + \alpha_{1000} + \alpha_{2000}}{4} \quad (1)$$

Thereby, NRC of PU foams was selected as target objective, the higher NRC the better sound absorbing. Upon completion of the nine experimental runs, the corresponding response, sound absorption coefficient tested by acoustical property measurements.

IOP Conf. Series: Materials Science and Engineering **824** (2020) 012016 doi:10.1088/1757-899X/824/1/012016

Table 2. Description of factors and their levels for Taguchi experimental design.

Variables	Parameter code	Level		
		-1	0	1
Monomer mixture	A	5	15	25
Filler ratio	B	1	3	5
Sample Thickness	C	10	20	30

Table 3. Orthogonal array L_9 (3^3) of experimental design.

Monomer	Filler Ratio	Sample Thickness
5	1	10
5	3	20
5	5	30
15	1	20
15	3	30
15	5	10
25	1	30
25	3	10
25	5	20

2.4. Sound Absorption Coefficient

Sound absorption test is performed to investigate the ability of certain material to absorb sound. The sound absorption coefficient was obtained by using a two-microphone impedance tube based on transfer function method in accordance to ASTM E1050-98 international standards (ASTM, 1998) [19]. Cylindrical samples with a diameter of 100 mm were used for the low-frequency range (100–2000 Hz) and samples with 28 mm diameter were used for the high-frequency range (200–5000 Hz). All the samples had a thickness of 10 mm, 20 mm and 30 mm. Figure 2 shows the sample foams used for acoustic test. Figure 3 presents the schematic for sound absorption test (a) low frequency tube test (b) high frequency tube test.

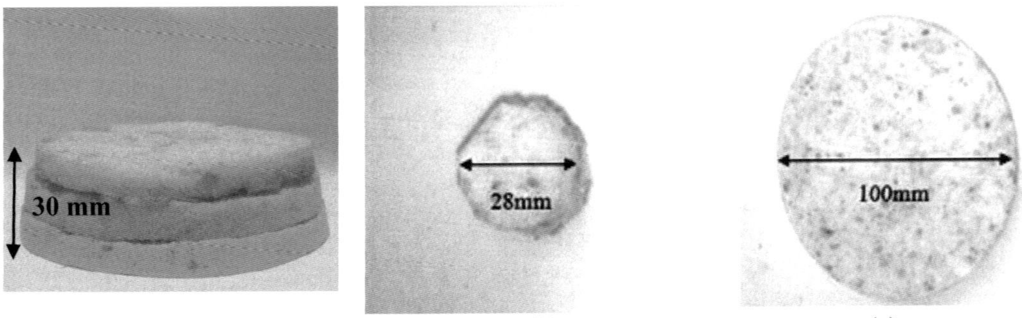

(a) (b) (c)

Figure 2. PUs foam (a) thickness 0f 10mm, 20mm and 30mm (b) diameter of 28 mm (c) diameter of 100 mm.

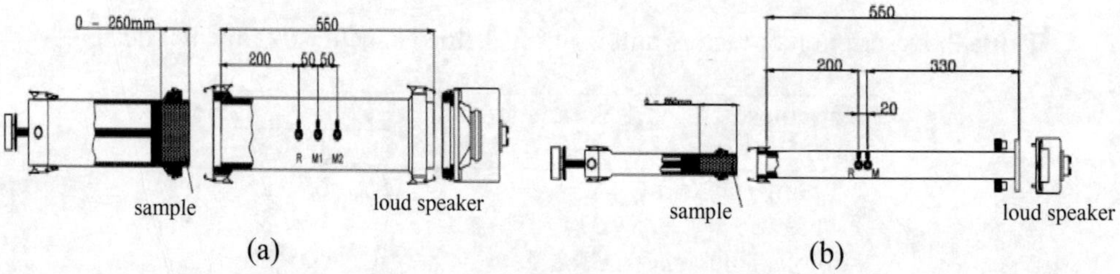

Figure 3. Schematic of the experimental setup for (a) low frequency (b) high frequency.

3. Result and Discussion

3.1. Signal-to-noise (S/N) ratio

Signal-to-noise ratio is used to measure the performance characteristic in Taguchi method, and then the experimental results are transformed to S/N ratio. Generally, there are three types of performance characteristic in the analysis of the S/N ratio, which are the lower the better, the higher the better and the nominal the better, respectively.

1) the lower the better:

$$\eta = -10\times\log\left(\frac{1}{n}\sum_{i=1}^{n} y_i^2\right) \tag{2}$$

2) the higher the better:

$$\eta = -10\times\log\left(\frac{1}{n}\sum_{i=1}^{n}\frac{1}{y_i^2}\right) \tag{3}$$

3) the nominal the better:

$$\eta = -10\times\log\left(\frac{1}{ns}\sum_{i=1}^{n} y_i^2\right) \tag{4}$$

where η represents the S/N ratio, y_i represents a quality measurement, n represents the total number of the tests, s represents the standard deviation. The quality characteristic in this study is NRC and the higher NRC is preferred. Therefore, the quality characteristic of the NRC is the higher the better.

3.2. Analysis of the S/N ratio

Based on the results of acoustic performances, the S/N ratio was used to determine the important effect of three parameters. The experiments results were further transferred to S/N ratio. According to the quality characteristic of the results, a higher S/N ratio corresponds to the better quality. The optimal level of three factors is the level with the highest S/N ratio. Table 4 shows the experiment results and corresponding S/N ratio for NRC. The S/N ratio for three levels of three factors for NRC is computed and listed in Table 5; the larger value of S/N ratio corresponds to larger NRC, which means the best sound absorption properties. The order in the Table 5 ranks the influence of the corresponding factors on the results.

The trend of the S/N ratios for three factors at their three designated levels affecting the sound absorption is shown in Figure 4. It is well known that the higher S/N ratio, the better the response for factorial effect. From the Figure 4, it can be clearly seen that the best combination of three factors to obtain the maximum NRC value for the S/N ratio is at level 3 of monomer mixture (25g bio-epoxy), level 2 of filler ratio content (3 wt%) and level 3 of sample thickness (30 mm). It can be also seen that the factor C (sample thickness 30 mm) has the most significant effect on the NRC, followed by factor A (monomer mixture) and factor B (filler ratio). So the optimal combination of parameters can be obtained as 'Monomer 25 - Filler ratio 3 - Sample Thickness 30' for NRC. The above results revealed that the sample thickness PU foam composites significantly affect the sound absorption coefficient. Therefore, suitable amount of monomer, filler ratio and sample thickness must be determined in order to obtain higher sound absorption of PU foam composites.

From this results, highest thickness of the PUs foam samples, the higher the sound absorption values [14]. Thus, the sound absorption has direct relationship with the thickness of the polymer foams. The increasing in thickness gives better performance of the materials as sound absorber which due to higher interconnected porous structure. This is also shows that the porous structure increased the level of vibration energy in the foam and gives highly versatile acoustic foam [20].

Table 4. Experimental results of NRC and corresponding S/N ratio.

Run	Factors			Response	
	A	B	C	NRC	S/N Ratio
1	5	1	10	0.12500	-18.0618
2	5	3	20	0.33250	-9.5642
3	5	5	30	0.46750	-6.6044
4	15	1	20	0.37000	-8.6360
5	15	3	30	0.48500	-6.2852
6	15	5	10	0.14750	-16.6242
7	25	1	30	0.46250	-6.6978
8	25	3	10	0.16975	-15.4038
9	25	5	20	0.36250	-8.81384

Table 5. S/N response for NRC.

Symbol	Factor	Level 1	Level 2	Level 3	Max-min	Order
A	Monomer	-11.410	-10.515	-10.305	1.105	2
B	Filler ratio	-11.132	-10.418	-10.681	0.714	3
C	Sample Thickness	-16.697	-9.005	-6.529	10.167	1

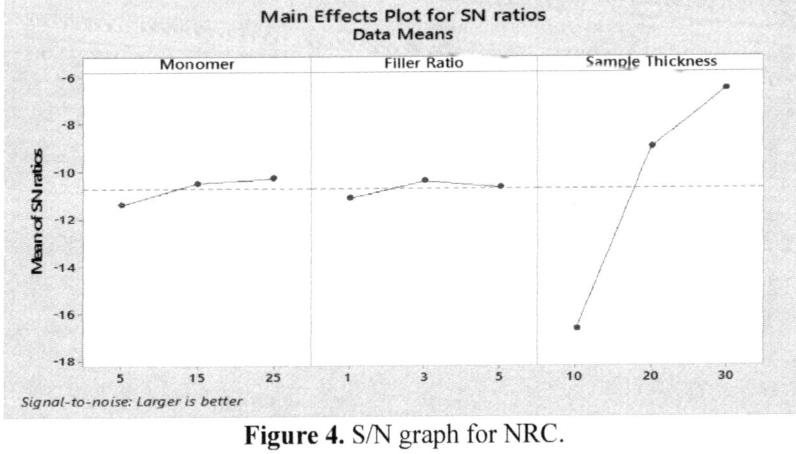

Figure 4. S/N graph for NRC.

3.3. Confirmation test

The optimum combination is not included in the orthogonal array $L_9(3^3)$, so it was necessary to conduct the confirmation test in order to validate the experimental results. The experiment was conducted making use of the above obtained optimum parameters. The curves for sound absorption coefficient of the material was shown in Figure 5 and compared with the initial sample points. The results showed the acoustic properties are improved effectively. By calculating the data obtained from the confirmation test (Table 6), the NRC of the optimal PU foams was 0.5550 and the highest sound absorption coefficient was up to 0.88. The NRC had showed an increased value from 0.4850 to 0.5550, improving 7.0% compared with Run 5 (Table 4), which showed best absorption properties among 9 runs.

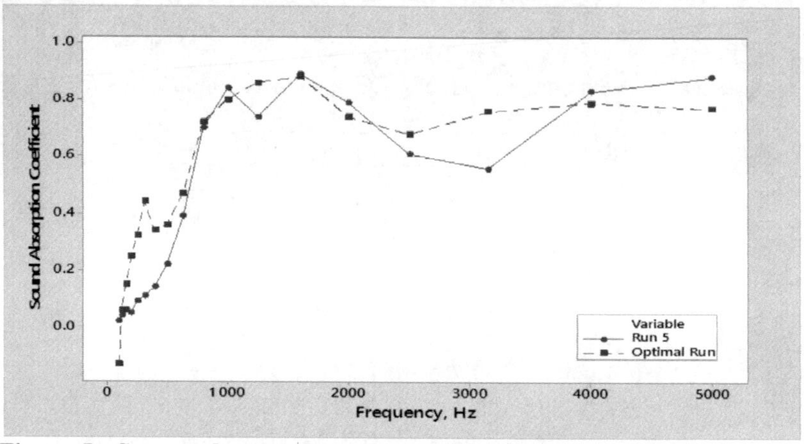

Figure 5. Curves of sound absorption coefficient for the optimal PU foams.

Table 6. The optimum design of formulation after optimization.

Parameters	A	B	C	NRC
The optimum design	25	3	30	0.5550

4. Conclusion

This study proposed an effective optimization approach to improve the acoustical behaviors of PU foam composites by optimizing the formulation of PU foams. Taguchi method based on the orthogonal array L_9 (3^3) was employed to conduct the experiments. Sound absorption coefficient was selected as the objective. To further achieve the optimum formulation of PU foams, three design variables were formulated aiming to simultaneously maximize the sound absorption. Signal-to noise (S/N) analysis was used to determine an optimal formulation of PU foam composites. The results showed that the acoustic properties of PU foam composites can be improved by optimizing the formulation of PU foams and the best combination values of variables were monomer mixture of 25 gram, filler ratio of 3% wt and sample thickness of 30 mm. By performing the confirmation test, the NRC had showed an increased value from 0.4850 to 0.5550 positively improved the acoustic property.

5. References

[1] S. Amares, E. Sujatmika, T. W. Hong, R. Durairaj, and H. S. H. B. Hamid, "A Review: Characteristics of Noise Absorption Material," in Journal of Physics: Conference Series, 2017.

[2] A. Z. M. Rus, M. S. M. Azahari, S. Kormin, L. B. Soon, M. T. Zaliran, and M. L. F. Ahraz Sadrina, "Hybrid waste filler filled bio-polymer foam composites for sound absorbent materials," in AIP Conference Proceedings, 2017.

[3] R. Gayathri and R. Vasanthakumari, "Nanomaterials in PU Foam for Enhanced Sound Absorption at Low Frequency Region," Adv. Mater. Res., 2014.

[4] J. G. Gwon, S. K. Kim, and J. H. Kim, "Sound absorption behavior of flexible polyurethane foams with distinct cellular structures," Quat. Int., 2015.

[5] M. N. A. A. Nordin, L. M. Wan, M. H. Zainulabidin, A. S. M. Kassim, and A. M. Aripin, "Research finding in natural fibers sound absorbing material," ARPN J. Eng. Appl. Sci., 2016.

[6] H. Abdul Latif, M. N. Yahya, M. N. Rafiq, M. Sambu, M. I. Ghazali, and M. N. Mohamed Hatta, "A Preliminary Study on Acoustical Performance of Oil Palm Mesocarp Natural Fiber," Appl. Mech. Mater., 2015.

[7] Y. Jiang, S. Chen, D. Wang, and J. Chen, "Multi-Objective Optimization of Acoustical Properties of PU-Bamboo-Chips Foam Composites," Arch. Acoust., 2017.

[8] M. S. M. Azahari, A. Z. M. Rus, M. Taufiq Zaliran, and S. Kormin, "Improving sound absorption property of polyurethane foams doped with natural fiber," IOP Conf. Ser. Mater. Sci. Eng., vol. 226, p. 012009, 2017.

[9] Abd Wahab, Hanani, Anika Zafiah Mohd Rus, & M. Taufiq Zaliran. "Design of Experiment for In-Situ Synthesis of Waste Vegetable Oils and Synthetic Based Polyurethane (PU) Foam Composites." International Journal of Engineering & Technology [Online], 7.4.34 (2018): 413-416. Web. 7 Mar. 2019

[10] R. Gayathri, R. Vasanthakumari, and C. Padmanabhan, "Sound absorption, thermal and mechanical behavior of polyurethane foam modified with nano silica, nano clay and crumb rubber fillers," Int. J. Sci. Eng. Res., vol. 4, no. 5, pp. 301–308, 2013.

[11] M. Kaseem, K. Hamad, F. Deri, and Y. G. Ko, "Material properties of polyethylene/wood composites: A review of recent works," Polym. Sci. Ser. A, 2015.

[12] P. M. Murray, F. Bellany, L. Benhamou, D. K. Bučar, A. B. Tabor, and T. D. Sheppard, "The application of design of experiments (DoE) reaction optimisation and solvent selection in the development of new synthetic chemistry," Organic and Biomolecular Chemistry. 2016.

[13] J. Y. Lau, M. H. T. Nguyen, W. Liang, and K. K. Tan, "Application of design of experiments to feature selection in ventilation tube applicator," in IECON 2015 - 41st Annual Conference of the IEEE Industrial Electronics Society, 2015.

[14] H. A. Wahab, A. Z. M. Rus, M. F. L Abdullah and N. M. Abdullah, "Design of Experiment for Sound Absorption Materials of Microporous Polymer," 2019 International Conference on Information Science and Communication Technology (ICISCT), Karachi, Pakistan, 2019.

[15] R. Davis and P. John, "Application of Taguchi-Based Design of Experiments for Industrial Chemical Processes," in Statistical Approaches With Emphasis on Design of Experiments Applied to Chemical Processes, 2018.

[16] H. Adeli, S. H. Sharif Zein, S. H. Tan and A. L. Ahmad, "Optimization of the mechanical strength properties of poly(L-lactide)/multi-walled carbon nanotube scaffolds using response surface methodology," Nano, 2011.

[17] K. L. Low, S. H. Tan, S. H. S. Zein, D. S. McPhail, and A. R. Boccaccini, "Optimization of the mechanical properties of calcium phosphate/multi-walled carbon nanotubes/bovine serum albumin composites using response surface methodology," Mater. Des., 2011.

[18] A. Z. M. Rus and N. N. Mat Hassan, "Elastic and Damping Characterization of Biopolymer Composite," Appl. Mech. Mater., vol. 799–800, pp. 187–191, 2015.

[19] ASTM International standard E-1050-98 (1998), Standard test method for impedance and absorption of acoustical materials using a tube, two micro-phones and a digital frequency analysis system, West Conshohocken, USA,1998.
[20] A. Z. M. Rus, "Sound Behavior Of Polymer Foam Composite With Different Composition Of Wood Dust Filler," in Conference proceeding MUCET2014 November, pp. 10–11, 2014.

Acknowledgements
The author would like to thank Sustainable Polymer Engineering, Advanced Manufacturing and Materials Center (SPEN-AMMC), Universiti Tun Hussein Onn Malaysia (UTHM). Also thanks to Aishah Sharkawi, Baie Dali, Shahfeeka Borhan, Sharifah Tuah, Zaki Shahmim, Anwar Adlin, Haziq Zaihan, Zulfadhli Mamat, Hafiz Nasri, Afnan Hamdan, Aminuddin Zaidi, Syakira Sarif, Nursyazwani Zulkefli, Haslinda Habe, Noorhafeez Yassin, Harry Leong, Betty Koh, Syafiq Norzari, Amin Rosnan, Aidil M.Noor, Hamdan Syukran, Sofreezan Soprudin and Ikram Amizan for completing this work.

IOP Conf. Series: Materials Science and Engineering **824** (2020) 012017 doi:10.1088/1757-899X/824/1/012017

Damage Initiation and Evolution Analysis of Hot Extruded Recycled Aluminium Alloys (AA6061)

C. S. Ho[1], M. K. Mohd Nor[1], N. Ma'at[1], K. Y. Alaric Sim[1], M. N. Ibrahim [1], S. Jamian[1], M. A. Lajis[2], N. K. Yusuf[2]

[1]Crashworthiness and Collisions Research Group, Mechanical Failure Prevention and Reliability Research Center (COLORED-MPROVE)
Faculty of Mechanical and Manufacturing Engineering,
Universiti Tun Hussein Onn Malaysia.

[2]Sustainable Manufacturing and Recycling Technology, Advanced Manufacturing, and Material Center (SMART-AMMC)
Faculty of Mechanical and Manufacturing Engineering,
Universiti Tun Hussein Onn Malaysia.

norihan@uthm.edu.my

Abstract. The effects of strain rate on the mechanical behaviour of recycled hot extruded aluminium alloys (AA6061) are investigated using uniaxial tensile tests in this work. The damage characteristics such as micro-cracks and micro-voids are analysed on the microstructure of the material using Scanning Microscope Electron (SEM). It is shown that the mechanical behaviour of the recycled AA6061 is influenced by the loading speed. Besides, the damage parameters such as the micro-cracks and micro-voids in the material are evolved significantly in such materials during finite strain deformation. It can be observed that the evolution speed of the micro-cracks and micro-voids in the materials is increasing with the increase in loading speed. Much work is still required to improve the recycled AA6061 as compared to the primary resources in terms of damage parameters initiation and evolution.

1. Introduction

The high demand of the primary aluminium alloys usage had been shifting the attention on the recycled aluminium alloys as the secondary resources. Nowadays, various recycling methods have been adopted in the production of recycled aluminium alloys capable of giving promising results. However, it is generally agreed that strain rates still a critical issue for consideration to ensure the identical engineering applications can be fulfilled as shown by the original solid-state form [1], [2]. At quasi-static rates of strain, this behaviour has been studied by a few researchers as published in [3], [4], [5] and [6]. Also, damage evolution in the recycled materials is still an existing open area to be studied in such materials.

Commercial aluminium alloys have been used extensively in aerospace and automotive industries due to its excellent profile such as low density, high specific modulus, excellent corrosion resistance and exhibits good machinability, formability and surface finish. Moreover, it also exhibits the potential of recyclability, where it can be fully and continuously recycled [5], [7]. However, during the production of aluminium alloys, it required a high amount of energy consumption, and it may bring to environment pollution [8], [9]. Therefore, a suitable replacement using recycled aluminium alloys as a secondary resource must be considered seriously.

The recycled aluminium alloys have produced from the aluminium scraps and chips. In previous, conventional technique that involved melting process is used to recycle the scrap and chips. However, some researchers found that the scraps and chips are not conducive and good for recycled using the re-melting process [10], [11]. Subsequently, solid-state recycling technique had been adopted. Figure 1 shows the comparison between the conventional and solid-state recycling technique to produce extruded recycled products. It can be noticed that the solid-state recycling technique required less procedure and less amount of new scrap will be produced in the recycling process [7], [12], [13].

However, there is a challenge on the recycled materials to achieve almost the same mechanical behaviour as the primary resources. This challenge due to the material degradation associated with the damage, such as the micro-crack and micro-voids in the recycled material. In fact, ductile fracture involves damage because of the nucleation, growth, and coalescence of those micro-cracks and micro-voids [14], [15].

In this paper, the recycle form of aluminium alloys (AA6061) was produced using a hot extrusion solid-state recycling technique. The specimens produced were tested at different strain rates via uniaxial tensile test, and the microstructure of the specimens was observed before and after the test. This approach allows for deep analysis on the damage behaviour of such materials undergoing finite strain deformation.

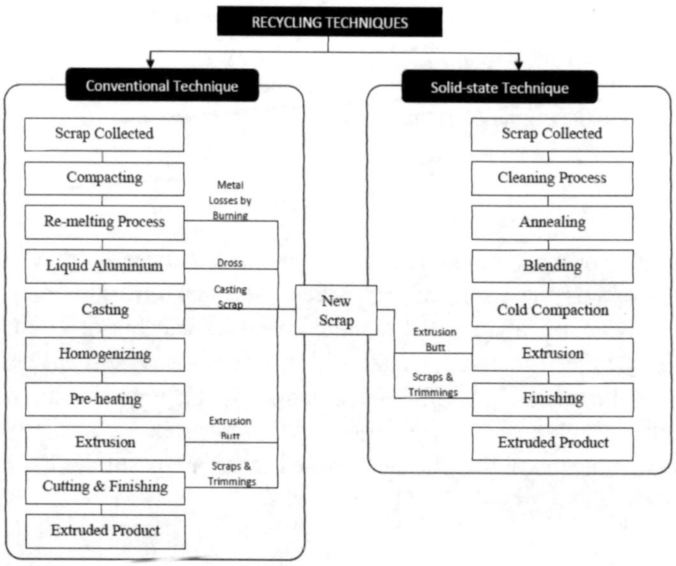

Figure 1. Flowchart of Conventional Recycling Technique and Solid-state Recycling Technique [12]

IOP Conf. Series: Materials Science and Engineering **824** (2020) 012017 doi:10.1088/1757-899X/824/1/012017

2. Experimental Procedure

2.1. Specimen Preparation

The aluminium alloys AA6061 chips were milled from AA6061-T6 plate by using Sodick-MC430L high-speed machine. Acetone solution was used to clean the chips by utilizing ultrasonic bath, and the cleaned chips were dried in the thermal oven at 60 °C for about 30 minutes. Next, cold press machine was used to form a 30 mm diameter with approximate 90 mm height cylindrical shape billet at 120 bar. Then, the billet was pre-heated in a 500 °C furnace for 120 minutes. Subsequently, the billet was inserted into thc 300 °C extrusion container and extruded. Lastly, the extruded product was cut into dog-bone shape according to ASTM E8 standard. The flow of the specimen preparation process is shown in Figure 2.

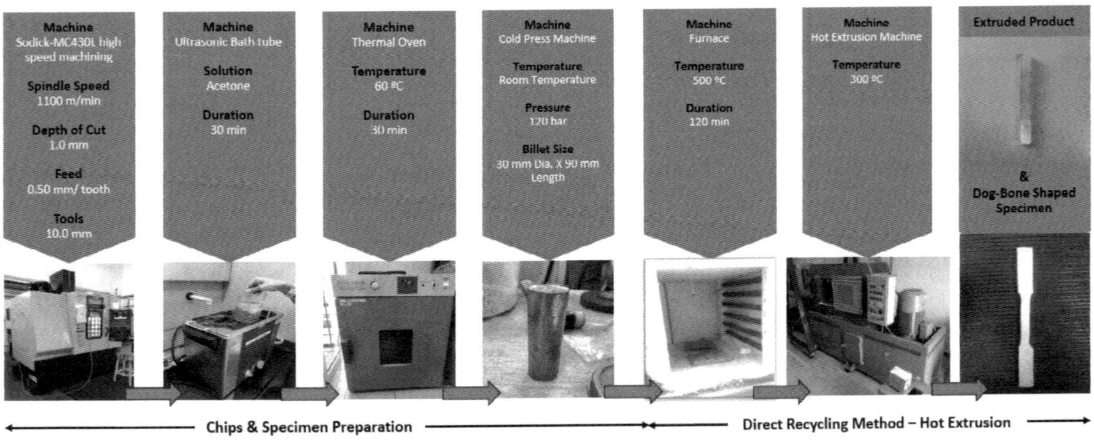

Figure 2. The flow of Specimen Preparation using Hot Extrusion

2.2. Uniaxial Tensile Test

The ASTM E8 dog-bone shaped recycled aluminium alloys (AA6061) specimens are tested in room temperature at three different loading speeds: 0.05 mm/min, 0.5 mm/min and 5 mm/min, using a uniaxial tensile machine. 25mm gauge length extensometer is used in order to obtain a more accurate result. The fracture surfaces after the tensile test arc then observed under SEM. The test matrixes of the experiment are shown in Table 1.

Table 1. Test Matrix of Uniaxial Tensile Test

Temperature :	Room Temperature
Specimen	Loading Speed (mm/min)
1	0.05
2	0.5
3	5

IOP Conf. Series: Materials Science and Engineering **824** (2020) 012017 doi:10.1088/1757-899X/824/1/012017

3. Results and Discussion

3.1. Mechanical Properties

Figure 3 and Table 2 presented the tensile results of the recycled AA6061 at different loading speed. The results of Young's modulus (E), yield strength and Ultimate Tensile Strength (UTS) are changing at different loading speed or strain rate. The UTS value is increasing; on the other hands, the performance of yield strength and Young's modulus are decreasing with the increasing of loading speed. The recycled AA6061 behave more elastic with the increase in loading speed. It can be noticed that the most sensitive parameter to the loading speed or strain rate is Young's modulus of the material, followed by UTS and yield strength.

Comparing with the primary resource, the mechanical performance of the hot extruded recycled AA6061 without heat treatment is poor than the as-received AA6061-T6, but it is still comparable with primary AA6061-T1. It can be noticed that the recycled AA6061 have some degradation on UTS value compared to AA6061-T1. The degradation of the mechanical performance might be caused by the micro-cracks and micro-voids within the recycled AA6061, which affected the bonding in the recycled material [16], [17]. The microstructure of the recycled material can be observed and analyzed in the following section.

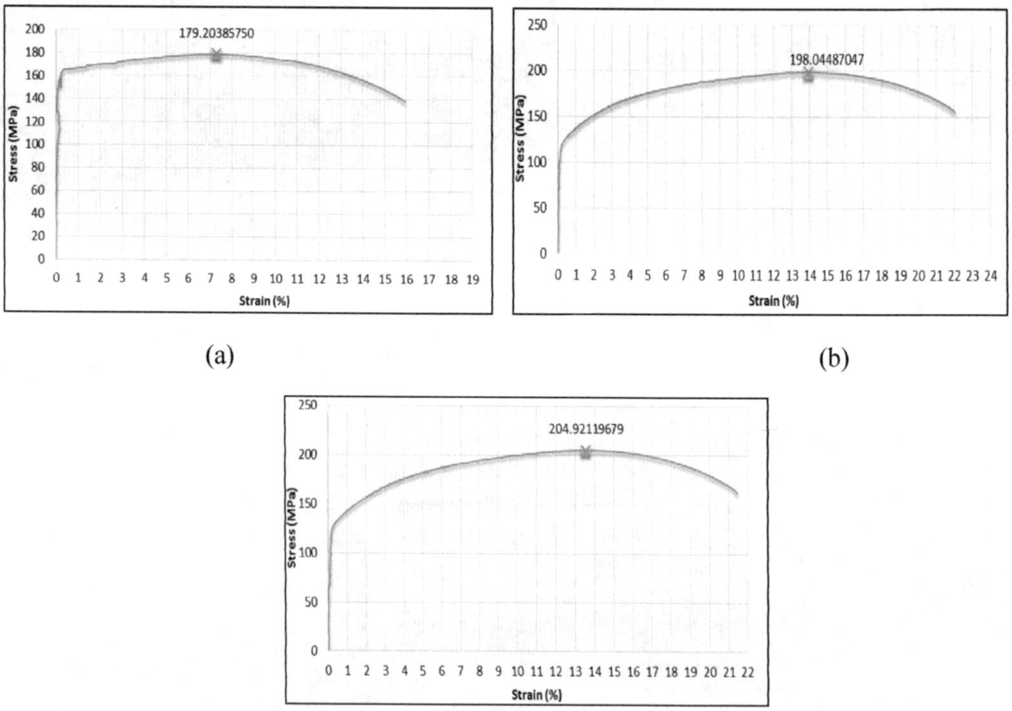

(a) (b) (c)

Figure 3. Tensile Test Results at Loading Speed of (a) 0.05 mm/min (b) 0.5 mm/min, (c) 5 mm/min

Table 2. Summary of Tensile Test Result

Specimen	Loading Speed (mm/min)	E (GPa)	Yield Strength (MPa)	UTS (MPa)
1	0.05	101.56	165.82	179.20
2	0.5	78.09	123.65	198.04
3	5	55.20	133.02	204.92
AA6061-T1 [18]	-	69	110	210
As-Received AA6061-T6	-	70	270	310

3.2. Microstructure

The tensile fracture surface of the recycled AA6061 was observed and analyzed by using SEM in this work, as illustrated in Figure 4. From the figure, the pores formed is the micro-voids and dimples in the fracture surface, and the lighter part which seems to be like strings is the crack propagation line [10], [19]. It can be observed that the amount of micro-voids and micro-cracks increase with increasing loading speed. At the highest loading speed, it can be observed that the voids and dimples are deeper while the quantity is higher than the one tested at lower loading speed.

In order to analyze the damage initiation and evolution, the microstructure before and after tensile loading is compared, as shown in Figure 5. Before undergoing tensile loading, the microstructure of the recycled AA6061 was smooth and fine. Less amount of micro-voids and micro-cracks can be noticed in the material. Contrarily, the formation of the micro-cracks and micro-voids due to the tensile loading is the main local damage initiation mechanism identified [20]. The micro-cracks and micro-voids evolved significantly in the recycled AA6061 undergoing tensile loading [21]. The evolution speed is increasing with the increment of loading speed or strain rate.

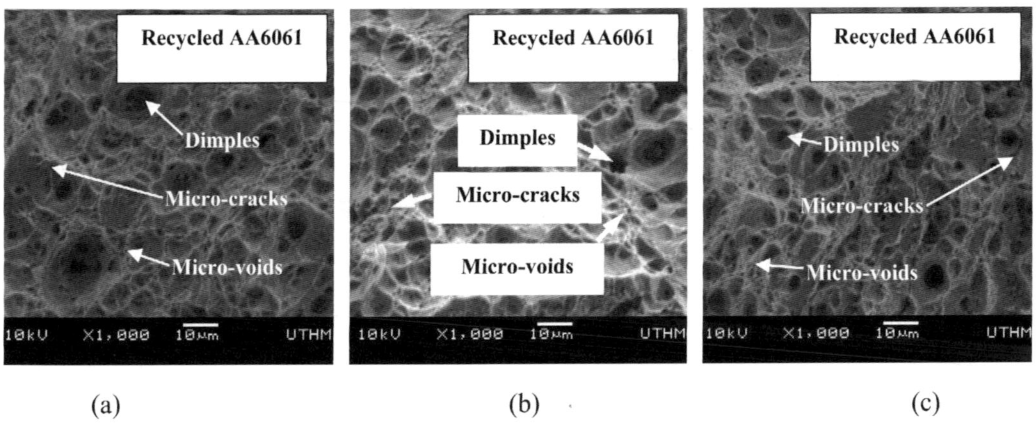

Figure 4. Microstructure of the Tensile Fracture Surface of Recycled AA6061 at Loading Speed of (a) 0.05 mm/min, (b) 0.5 mm/min, (c) 5 mm/min

IOP Conf. Series: Materials Science and Engineering **824** (2020) 012017 doi:10.1088/1757-899X/824/1/012017

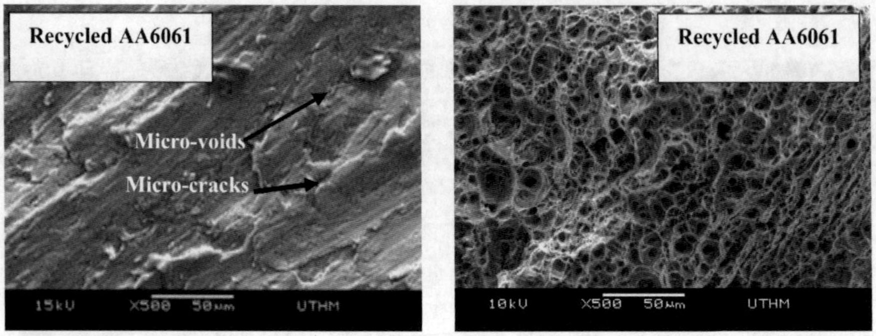

(a) (b)

Figure 5. Microstructure of (a) Undeformed and (b) Deformed Recycled AA6061

Further, the results of the recycled AA6061 is compared to the primary resource. Figure 6 shows the comparison of the microstructure between the hot extruded recycled AA6061 and the as-received AA6061. Numerous dimples and micro-voids can be seen in both the fracture surface of the materials, and the fracture mode is categorized as ductile fracture deformation [10]. Undergoing the same loading speed condition at 0.5 mm/min, as-received AA6061 shows better results than the recycled material in terms of damage parameters development and evolution. The amount of micro-cracks and micro-voids in the primary resource is lesser than the recycled form, and the size of those damages can be seen smaller too. A lot of deep dimples and micro-voids can be found in the recycled AA6061 compared to the primary resource.

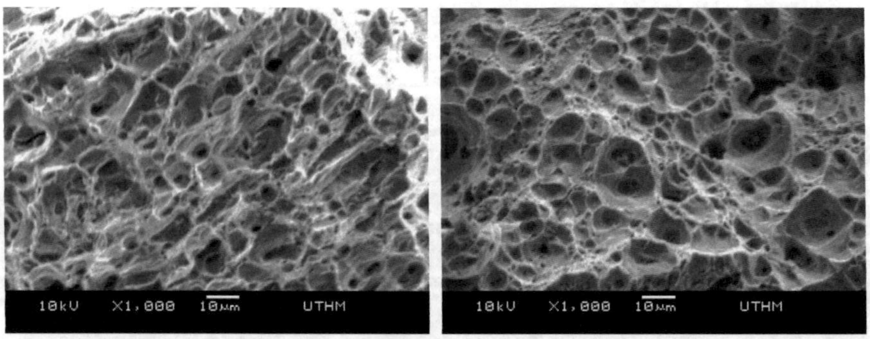

Figure 6. Comparison between the Microstructure of Deformed Primary Aluminium Alloys and Recycled Aluminium Alloys Specimen

4. Conclusion

The results in this work show that the mechanical behaviour of the selected recycled material is dependent on the given loading speed as generally accepted. The material can withstand higher UTS value and behaves elastically with the increasing of loading speed. Based on the micrographs analysis, the microstructure of the undeformed hot extruded recycled AA6061 looks smooth, and not much micro-voids and micro-cracks can be found. The damage parameters start to evolve due to the growth and coalescence of the voids and cracks. Moreover, it can also be observed that the loading speed influences the damage evolution. The voids and dimples that developed in the deformed recycled AA6061 at highest deformation speed are deeper and much higher than the one at lower loading speed.

In summary, much work still needed to enhance the capability of the recycled AA6061 as fulfilled by the primary resource. It is due to the damage parameters developed inside the recycled materials as proven in this work. In future works, heat treatment and aging are suggested to be included in the development process of the recycled material to improve the bonding between microstructures inside the recycled material.

5. Acknowledgement

Authors wish to convey sincere gratitude to Universiti Tun Hussein Onn Malaysia (UTHM) for providing the financial means during the preparation to complete this work under Geran Penyelidikan Pascasiswazah (GPPS), Vot U746 and UTHM Contract Research Grant, Vot H276.

References

[1] M. K. Mohd Nor, "Modelling inelastic behaviour of orthotropic metals in a unique alignmentof deviatoric plane within the stress space," *Int. J. Non. Linear. Mech.*, vol. 87, no. February,pp. 43–57, 2016.

[2] M. K. M. Nor, N. Ma'at, and C. S. Ho, "An Anisotropic Elastoplastic Constitutive Formulation Generalised for Orthotropic Materials," *Contin. Mech. Thermodyn.*, vol. 30, no. 7, pp. 1–36, 2018.

[3] R. Vignjevic, N. K. Bourne, J. C. F. Millett, and T. De Vuyst, "Effects of orientation on the strength of the aluminum alloy 7010-T6 during shock loading: Experiment and simulation," *J. Appl. Phys.*, vol. 92, no. 8, pp. 4342–4348, 2002.

[4] S. Sinha and S. Ghosh, "Modeling cyclic ratcheting based fatigue life of HSLA steels using crystal plasticity FEM simulations and experiments," *Int. J. Fatigue*, vol. 28, no. 12, pp. 1690–1704, 2006.

[5] N. Ma'at, M. K. M. Nor, C. S. Ho, N. A. Latif, A. E. Ismail, K.-A. Kamarudin, S. Jamian, M. N. Ibrahim@Tamrin, and M. K. Awang, "Effects of Temperatures and Strain Rate on the Mechanical Behaviour of Commercial Aluminium Alloy AA6061," *J. Adv. Res. Fluid Mech. Therm. Sci.*, vol. 54, no. 1, pp. 21–26, 2019.

[6] M. K. M. Nor and I. M. Suhaimi, "Effects of Temperature and Strain Rate on Commercial Aluminum Alloy AA5083," *Appl. Mech. Mater.*, vol. 660, pp. 332–336, 2016.

[7] S. N. A. Rahim, M. A. Lajis, and S. Ariffin, "A review on recycling aluminum chips by hot extrusion process," *Procedia CIRP*, vol. 26, no. September 2016, pp. 761–766, 2015.

[8] J. Cui and H. J. Roven, "Recycling of automotive aluminum," *Trans. Nonferrous Met. Soc. China (English Ed.*, vol. 20, no. 11, pp. 2057–2063, 2010.

[9] G. Gaustad, E. Olivetti, and R. Kirchain, "Improving aluminum recycling: A survey of sorting and impurity removal technologies," *Resour. Conserv. Recycl.*, vol. 58, pp. 79–87, 2012.

[10] N. K. Yusuf, M. A. Lajis, and A. Ahmad, "Hot Press as a Sustainable Direct Recycling Technique of Aluminium : Mechanical Properties and Surface Integrity," *Materials (Basel).*, vol. 10, no. 8, p. 902, 2017.

[11] J. Z. Gronostajski, H. Marciniak, and A. Matuszak, "Production of composites on the base of AlCu4 alloy chips," *J. Mater. Process. Technol.*, vol. 60, pp. 719–722, 1996.

[12] M. Samuel, "A new technique for recycling aluminium scrap," *J. Mater. Process. Technol.*, vol. 135, no. 1, pp. 117–124, 2003.

[13] A. Ahmad, M. A. Lajis, N. K. Yusuf, S. Shamsudin, and Z. W. Zhong, "Parametric Optimisation of Heat Treated Recycling Aluminium (AA6061) by Response Surface Methodology," in *AIP Conference Proceedings*, 2017, vol. 1885, no. 1, pp. 1–7.

[14] G. Z. Voyiadjis and P. I. Kattan, *Damage Mechanics*. CRP Press, 2005.

[15] C. S. Ho, M. A. A. Rani, M. K. M. Nor, N. Ma'at, M. T. H. Sultan, M. A. Lajis, and N. K. Yusuf, "Characterization of Anisotropic Damage Behaviour of Recycled Aluminium Alloys AA6061 Undergoing High Velocity Impact," *Int. J. Integr. Eng.*, vol. 11, no. 1, pp. 247–256, 2019.

[16] B. C. Ray, "Effects of Changing Environment and Loading Speed on Mechanical Behavior of FRP Composites," *J. Reinf. Plast. Compos.*, vol. 25, no. 12, pp. 1227–1240, 2006.

[17] R. Chiba and M. Yoshimura, "Solid-state recycling of aluminium alloy swarf into c-channel by hot extrusion," *J. Manuf. Process.*, vol. 17, pp. 1–8, 2015.

[18] MakeItFrom.com, "6061-T1 Aluminum," *6061 Aluminum in Depth*, 2017. [Online]. Available: https://www.makeitfrom.com/material-properties/606.

[19] V. Güley, A. Güzel, A. Jäger, N. Ben Khalifa, A. E. Tekkaya, and W. Z. Misiolek, "Effect of die design on the welding quality during solid state recycling of AA6060 chips by hot extrusion," *Mater. Sci. Eng. A*, vol. 574, pp. 163–175, 2013.

[20] H. Ghadbeigi, C. Pinna, and S. Celotto, "Failure mechanisms in DP600 steel: Initiation, evolution and fracture," *Mater. Sci. Eng. A*, vol. 588, pp. 420–431, 2013.

[21] C. Apostolopoulos, A. Drakakaki, A. Apostolopoulos, T. Matikas, A. I. Rudskoi, and G. Kodzhaspirov, "Characteristic defects-corrosion damage and mechanical behavior of dual phase rebar," *Mater. Phys. Mech.*, vol. 30, no. 1, pp. 1–19, 2017.

Descriptive statistics analysis of the variable in the data of toothbrushing simulator system modelling

Salihatun Md Salleh[1],*, Mohamad Norani Mansor[1], Hadirah Hassan[2], Ainul Husna Mohd Yusoff[1], Badrul Aisham Md Zin[1], Musli Nizam Yahya[1], S. T. Wijianto[3]

[1]Mechanical Failure and Reliability (MRPOVE), Department of Engineering Mechanics, Faculty of Mechanical and Manufacturing Engineering, Universiti Tun Hussein Onn Malaysia (UTHM), Batu Pahat 86400 Johor, Malaysia.
[2]Faculty of Computer and Mathematical Sciences, Universiti Teknologi MARA (UiTM), Seremban 3, 70300, Negeri Sembilan, Malaysia.
[3]Faculty of Mechanical Engineering, Universitas Muhammadiyah Surakarta, Surakarta 57162, Jawa Tengah, Indonesia.

E-mail: saliha@uthm.edu.my

Abstract. The controller of a system can be designed by constructing the model from the known the input and output. Modelling the dynamic of the system can help to visualize the real behavior of the system to develop the suitable controller parameter. The measured data from the real system need to be analyzed statistically to support that the current system needs a controller to improve the system performance. This study analyzed the normality and linearity of the variables in the research data of Toothbrushing Simulator System Modelling which are Speed (RPM) as the output(Y) and Voltage (V) as the input (X). There were five different data sets with 1000 observations respectively. All the data had been analyzed by using IBM SPSS Statistics 23 in which it will be explained by graphical method of histogram, scatter plot and descriptive measures of coefficient f determination between variables. The results for this study turn out that all the data sets were not normally distributed and not linear. Hence, the result from this statistical analysis has proven that the controller development is very crucial for the toothbrushing simulator system to improvise the system performance.

1. Introduction

Statistically, it is conventional to assume that the observations or samples collected in a research are normally distributed. Normality is often a maintained assumption in estimation and finite-sample inference [1]. The assumption of normality is useful as significant test of the models and to examine the confidence intervals around parameters [2]. The normality distribution can be examined by using graphical method and numerical method [3]. Compared to numerical method, graphical methods are easy to interpret and intuitive as they visualize the distribution of random variables. Whereas, numerical methods provide objective ways of examining normality as its present summary statistics included skewness, kurtosis and statistical test of normality [4].
Linearity is the ability within a given range to give out the results that are directly proportional to the concentration of the analyte in the test sample [5]. Linearity also can be defined when the input (X) and corresponding output (Y) are tied by the rules of superposition and homogeneity, then the device

or system behavior is said to be linear, otherwise nonlinear [6]. The relationship between two variables should always be made on the basis of a scatter plot that will determine the linearity [7].
The purpose of this study is to analyses the normality and linearity of the variables in the research data of Toothbrushing Simulator System Modelling which are Speed (RPM) as the output(Y) and Voltage (V) as the input(X). This study is primarily concerned on the normality and linearity test of the real time input and output signal of the system to support that the current system needs a controller to improve the system performance.

2. Methodology

2.1 Sample

Finite The purpose of this statistical analysis is to prove that the controller development is very crucial for the toothbrushing simulator. Hence, this study used the research data of Toothbrushing Simulator test rig [8],[9]. There were 1000 samples in each of five different data sets of Pulse Width Modulation (PWM) respectively which were PWM 175, 195, 215, 235 and 255. Whereby each data set consist of 1000 samples of real time input and output data. All input and output signals involved are represented in Fig. 1 and 2.

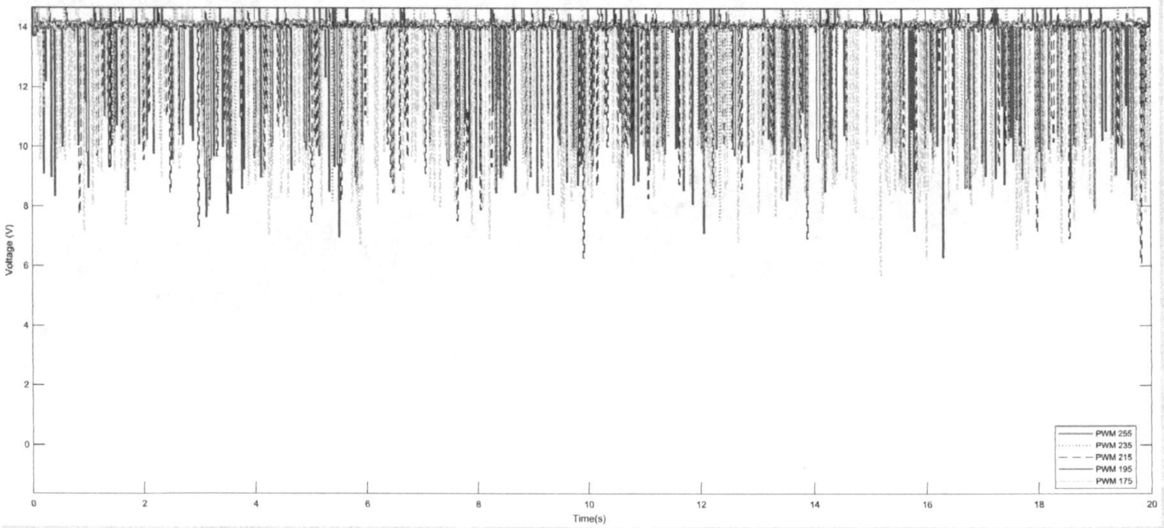

Figure 1. Input signal of five data set of PWM 255, 235, 215, 195 and 175.

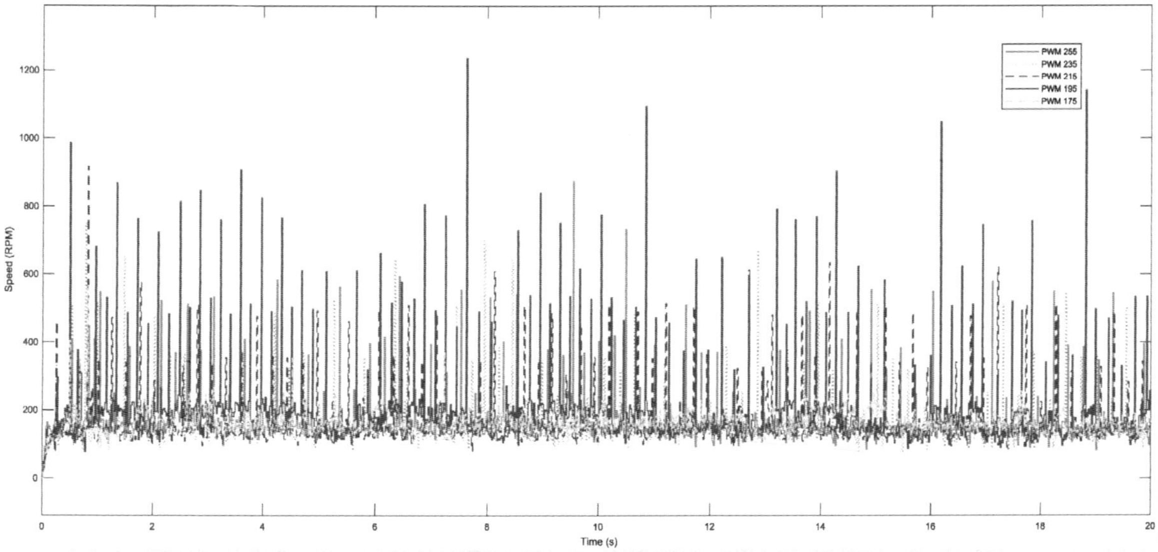

Figure 2. Output signal of five data set of PWM 255, 235, 215, 195 and 175.

2.2 Normality and Linearity Test

It is interesting to study the distribution of the data referring to the arithmetic mean to characterizing the dynamics of the average. Naive intuition suggests that most data will be randomly distributed about some preferred set of values when the system is in a steady state, as there are many ways in which data can be randomly distributed. For normality, plotting a histogram of the variable of interest is one of the graphical methods for checking the normality of data in which the observed value is plotted against their frequency [10]. The ideal appearance type which is a bell-shape curve will give an indication that the data is normally distributed [11].

In addition, normality test using skewness and kurtosis is another method in accessing the normality of the distribution as it may be more reliable for small and large samples. A measure of asymmetry of the distribution of variables is called as skewness while kurtosis is a measure of the peakedness of a distribution [12], [13]. The SPSS Software will give the value of skewness and kurtosis along with the value of standard error. The formula used in [14] to measure the standard error, SE of skewness and kurtosis are given by equation (2) below.

$$SE\ of\ Skewness = \sqrt{\frac{6}{n}},\ SE\ of\ Kurtosis = \sqrt{\frac{24}{n}} \qquad (1)$$

where n is the sample size.

Meanwhile, a scatter plot will help in examining the linearity for a data set by plotting the graph of input at x axis versus output at y axis [7]. A scatter plot is a simple method in identifying the relationship between two variables of input(X) and output(Y) [15]. The linear equation can be represented as in equation (2).

$$y = mx + c \qquad (2)$$

where m is the slope and c are constant. In order to get the linear relationship, as stated in [6], c value must be zero. All the analysis had been analysed by using a software of IBM SPSS Statistics 23 [16].

IOP Conf. Series: Materials Science and Engineering **824** (2020) 012018 doi:10.1088/1757-899X/824/1/012018

3. Result and Discussion

3.1 Normality

A common pattern which is the bell-shaped curve with the zero skew value will indicate the data is normally distributed. The descriptive statistic of skewness value for both variables in all PWM had been presented in Table 1. Standard error was written as 0.077 as this value is consistent at all reading. The skewness for both variables of all the dataset can be seen through the histogram plotted using SPSS in the Table 2. The skewness shows that the distribution is not in normal in most conditions. Overall, the PWM of speed and voltage variables were positive skew value as it was skewed to the right.

However, PWM 255 shown the voltage had a negative skew, as it has left skewed. Table 2 illustrates the distribution of variables in all PWM. Most of the sample mean values were clustered on the left side for speed variable. While, there was a spike plotted in each histogram of voltage as it represents the most common values except for PWM 255 that shown the skewness was heavily to the left. It shows that the data was not normally distributed.

Table 1. Descriptive Statistics.

Data Set	Skewness	Variables	
		Speed	Voltage
175	Statistic	2.313	1.410
	Std. Error	0.077	0.077
195	Statistic	3.547	1.814
	Std. Error	0.077	0.077
215	Statistic	3.879	2.170
	Std. Error	0.077	0.077
235	Statistic	4.292	3.395
	Std. Error	0.077	0.077
255	Statistic	4.110	-9.214
	Std. Error	0.077	0.077

Table 2. Histogram with Normal Curve Plotted of Speed (RPM) and Voltage(V).

<table>
<tr><th>PWM</th><th>Speed (RPM)</th><th>Voltage (V)</th></tr>
<tr><td>175</td><td>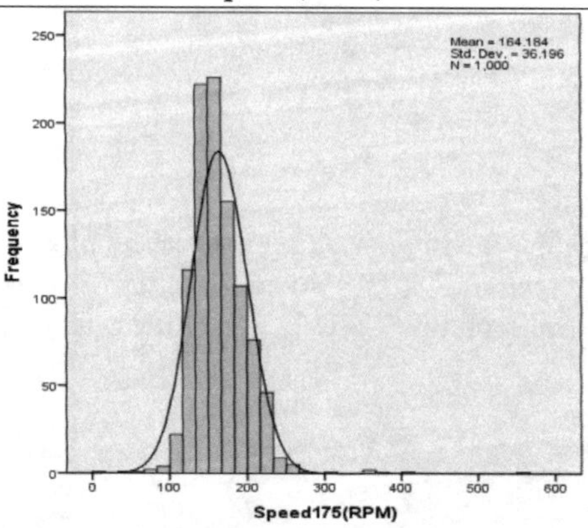
</td><td>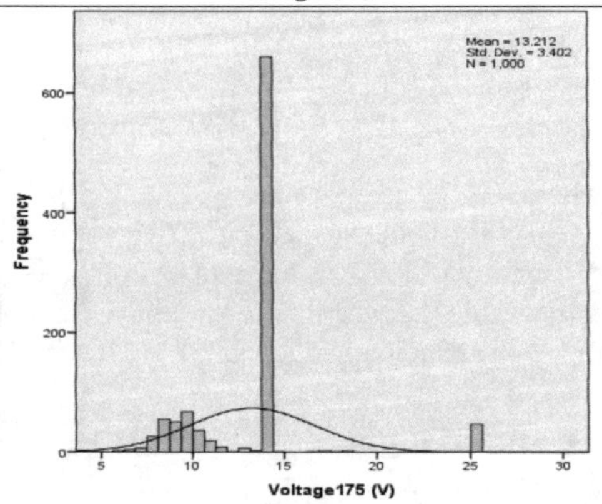
</td></tr>
</table>

IOP Conf. Series: Materials Science and Engineering **824** (2020) 012018 doi:10.1088/1757-899X/824/1/012018

195

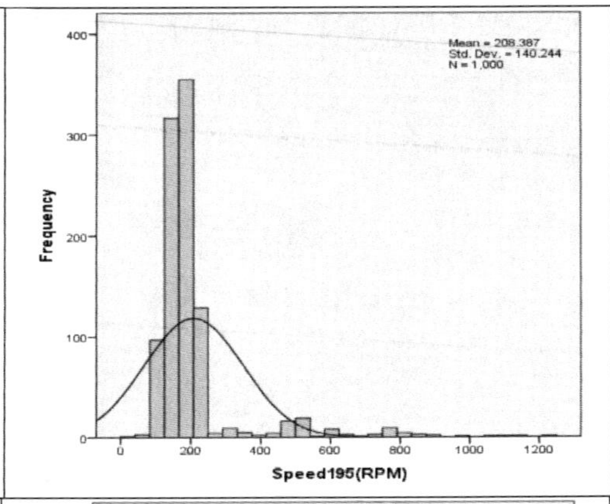

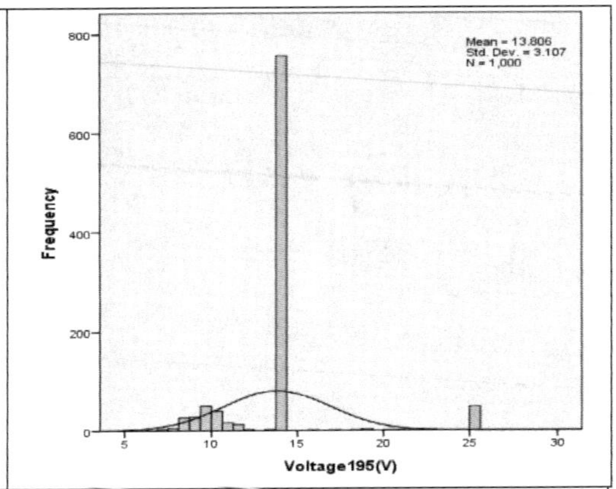

215

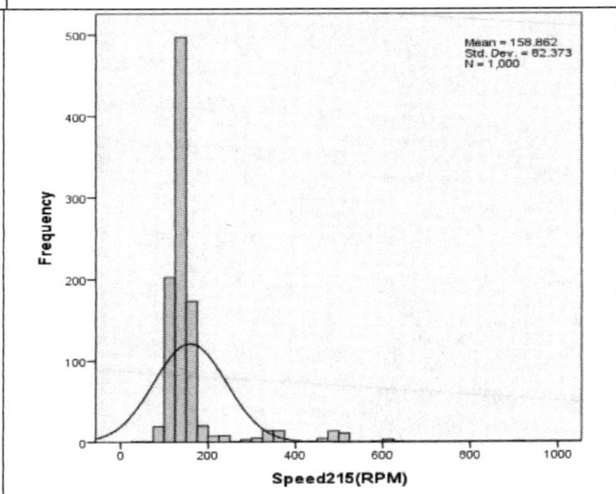

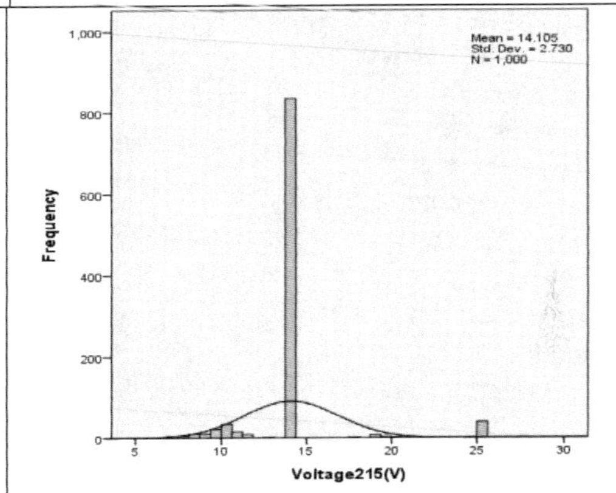

235

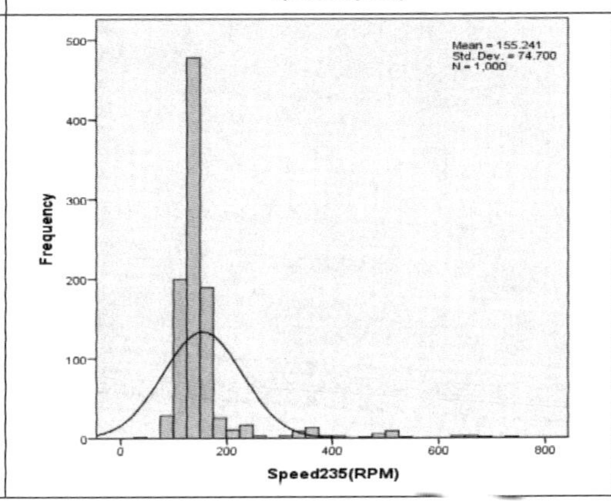

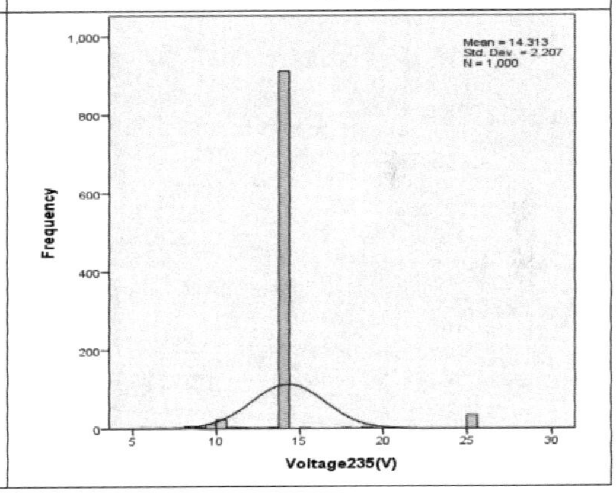

IOP Conf. Series: Materials Science and Engineering **824** (2020) 012018 doi:10.1088/1757-899X/824/1/012018

(Cont.) Table 2: Histogram of Speed (RPM) and Voltage(V).

PWM	Speed (RPM)	Voltage (V)
255	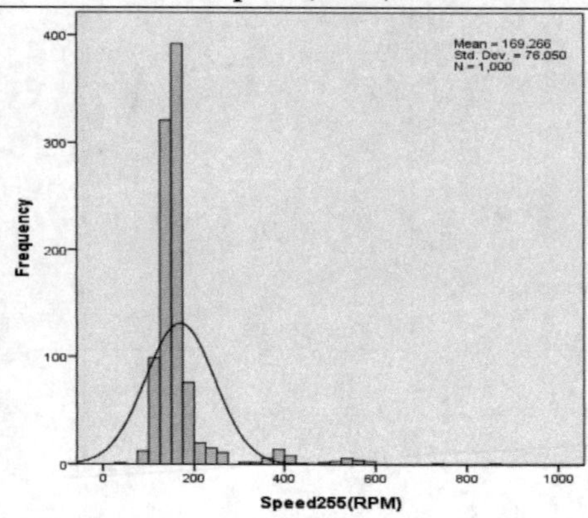	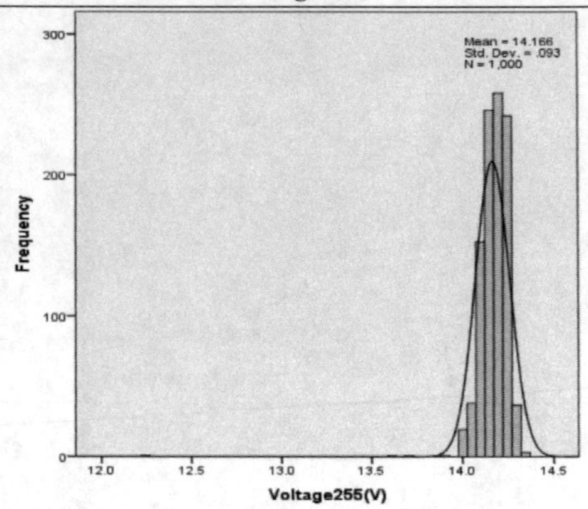

3.2 Linearity

Based on Fig. 3 to 7, they visualize that the plotting are not randomly scatted. Thus, it can be concluded that there was no linear relationship between speed and voltage in all dataset of PWM since there were a random pattern in the graph plotted. In addition, these also can be proven by checking the statistical measure, which is coefficient of determination, R^2 that will measure the strength of the linear relationship between the dependent and independent variables

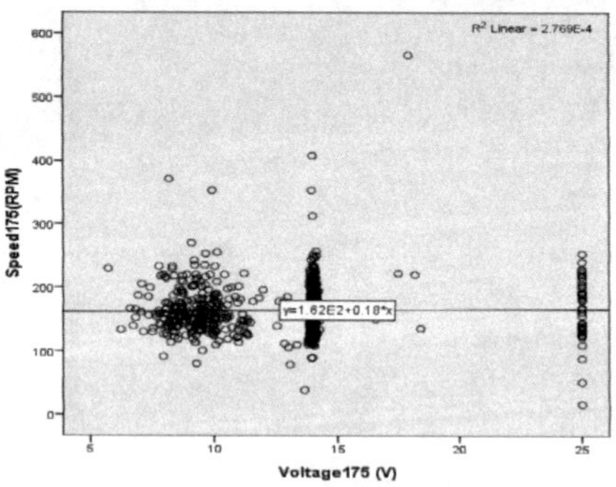

Figure 3. Linearity of Speed (RPM) and Voltage(V) in PWM175.

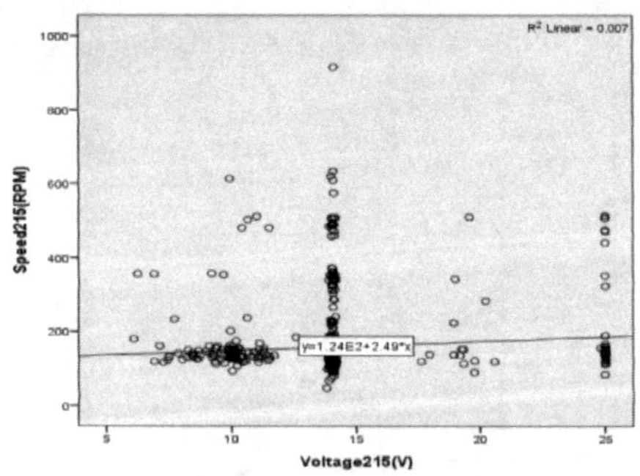

Figure 5. Linearity of Speed (RPM) and Voltage(V) in PWM215

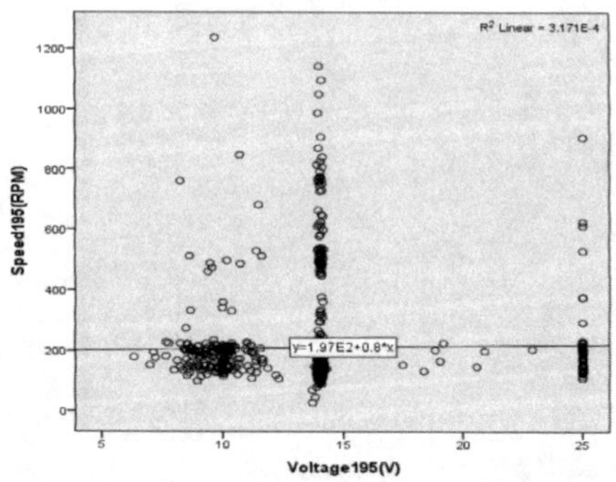

Figure 4. Linearity of Speed (RPM) and Voltage(V) in PWM195.

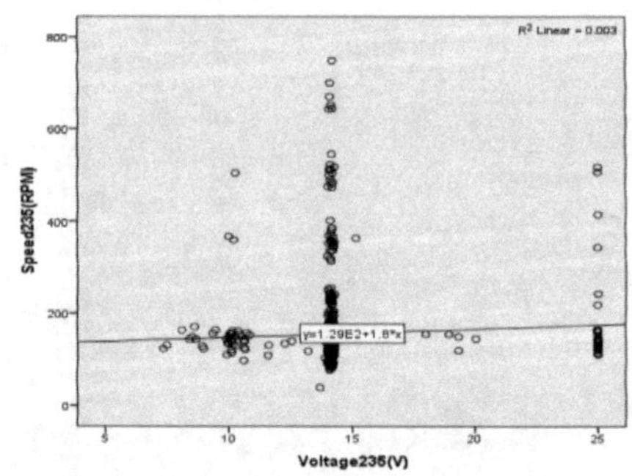

Figure 6. Linearity of Speed (RPM) and Voltage(V) in PWM235.

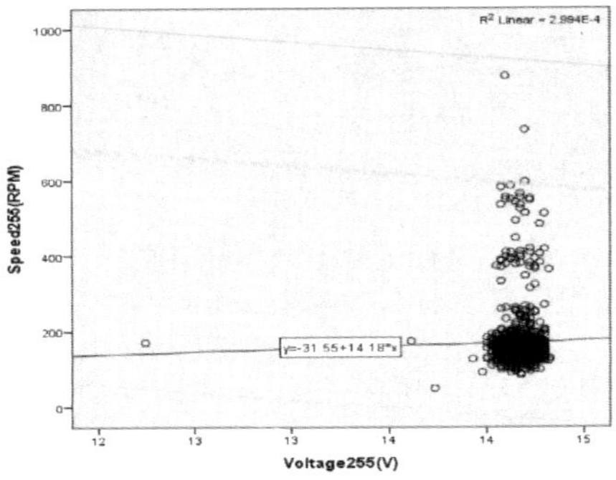

Figure 7. Linearity of Speed (RPM) and Voltage(V) in PWM255.

Table 3. Linearity of Speed (RPM) and Voltage(V).

PWM	Coefficent of Determination, R^2	Conclusion
175	0.051	Non-linear
195	0.058	Non-linear
215	0.007	Non-linear
235	0.003	Non-linear
255	0.055	Non-linear

4. Conclusion

According to the results, all the data sets can be concluded that the distribution of both variables, input and output are not normally distributed. As for linearity, the figures of scatter plot gave evidence that the association between input, voltage and output, speed variables is not linear. It is also proven that the system is a nonlinear. Adding the sample of observations is recommended for further study to imply the normal distribution and non inear system. In conclusion, the proposed testing can be used to determine the system condition to proof that the controller development is very crucial for the toothbrushing simulator system to improvise the system performance.

Acknowledgement

The authors graciously acknowledge the financial support provided by Universiti Tun Hussein Onn Malaysia, UTHM.

References

[1] Bai, J., & Ng, S. (2005) Tests for Skewness, Kurtosis, and Normality for Time Series Data. Journal of Business & Economic Statistics, 49-60.

[2] Field, P. A. (2016). Discovering Statistic, 1-20.

[3] Lund, D. A., & Lund, M. (2015). Laerd Statistics. Retrieved 25 October 2018, from Laerd Statistics: https://statistics.laerd.com/spss-tutorials/testing-for-normality using-spss-statistics.php

[4] Park, H. M. (2008). Univariate Analysis and Normality Test Using SAS, Stata, and SPSS. University Information Technology Services Center for Statistical and Mathematical Computing Indiana University, 1-38.

[5] LGC Maine Standards. (2017). Retrieved 24 October 2018, from LGC: https://www.mainestandards.com/support/faq-calibration-verification.php.

[6] Electrical System. (September, 2010). Retrieved 24 October 2018, from Electrical System: http://www.skm-eleksys.com/2010/09/linear-and-non-linear-systems-in.html

[7] Schneider, A., Hommel, G., & Blettner, M. (2010). Linear Regression Analysis. Series on Evaluation of Scientific Publications, 776-781.

IOP Conf. Series: Materials Science and Engineering **824** (2020) 012018 doi:10.1088/1757-899X/824/1/012018

[8] Ainul H.M.Y, Salleh S. M., Halib N., Taib H., Fathi, M.S. (2018). Analysis of Toothbrush Rig Parameter Estimation Using Different Model Orders in Real-Coded Genetic Algorithm (RCGA). International Journal of Engineering & Technology (IJET), 7, 443-447.
[9] Yusoff A.H.Y., Salleh S.M., Mahzan S., Ahmad S., Taib A., Rahman H.A., Ismail A., Choiron M.A. (2019). Evaluation of Varied Model Order in GA-optimised Parameter Estimation of Toothbrush Rig System. International Journal of Integrated Engineering.
[10] Ghasemi, A., & Zahediasl, S. (2012). Normality Tests for Statistical Analysis: A Guide for Non-Statisticians. International Journal of Endocrinology & Metabolism, 487-489.
[11] Das, K. R., & Imon, A. R. (2016). A Brief Review of Test for Normality. American Journal of Theoritical and Applied Statistics, 5-12.
[12] Kim, H. Y. (2013). Statistical notes for clinical researchers: assessing normal distribution (2) using skewness and kurtosis. *Restorative Dentistry & Endodontics*, 52-54.
[13] Rose, S., Spinks, N., & Canhoto, A. I. (2015). Management Research: Applying the Principle. Test for The Assumption That A Variable Is Normally Distributed, 1-4.
[14] Shukla, Satishprakash, S. (2015). Normality Test. Conference in New Trend in Educational Research, Department of Education, Gujerat University, Ahmedabad, Gujerat, India.
[15] Dastmard, B. (2013). A statistical analysis of the connection between test results and field claims for ECUs in vehicles. Stockholm, Sweden: Royal Institute of Technology School of Engineering Sciences.

IOP Conf. Series: Materials Science and Engineering **824** (2020) 012019 doi:10.1088/1757-899X/824/1/012019

Characterization of Continuous Gradient Functionally Graded Natural Fiber Reinforced Polymer Composites

M I H Jamaludin,[1] S Jamian,[1,2,3] M K Awang,[1,2] K A Kamarudin,[1,2] M K Mohd Nor[1,2] and A E Ismail[1,2]

[1]Department of Mechanics Engineering, Faculty of Mechanical and Manufacturing Engineering, Universiti Tun Hussein Onn Malaysia, 86400 Parit Raja, Johor, MALAYSIA.

[2]Crashworthiness and Collisions Research Group, Faculty of Mechanical and Manufacturing Engineering, Universiti Tun Hussein Onn Malaysia, 86400 Parit Raja, Johor, MALAYSIA.

E-mail : saifulnz@uthm.edu.my

Abstract. In this study, cylindrical tubes of functionally graded natural fibre reinforced polymer (FGNF/epoxy) composite are fabricated using horizontal centrifugal casting method. Natural fibres (NF) used in this research are coconut coir and oil palm empty fruit bunch (OPEFB). Natural fibres are mixed with epoxy at four different compositions which is at 0%, 5%, 10% and 15%. The study aims to determine the mechanical properties and microstructure of the FGNF/Epoxy composites. In this method, a homogenous mixture of epoxy liquid and natural fibres is inserted into mould simultaneously. Then, the mould is rotated at a constant speed for 4 hours. Effect of this process, the natural fibres particles in epoxy liquid move radial during mould rotation. Hardness and compression test are carried out to characterize the FGNF/epoxy cylindrical tubes. The microstructures of the composites are observed using optical microscope (OM) and scanning electron microscope (SEM). From the microstructural observation, it is found that the natural fibres particles can be graded from inner to outer surfaces of the FGNF/epoxy cylindrical tubes. The hardness value of the outer part of the FGNF/epoxy fabricated is higher than the middle and inner parts. Besides, from the compression test it shows that epoxy reinforced NF give a better result in strength compared to specimen without reinforced NF. However, the strength value decrease when the fibres percentage increase more than 5%. FGNF/epoxy with 5% NF content gives a balanced composition to have high hardness value and strength.

1. Introduction

Nowadays, natural fibre reinforced polymer (NF/P) composites are increasingly being used for a wide range of applications due to many of their advantages such as in automotive and industrial sector. Natural fibres have the advantages of low density, low cost recyclability and biodegradability. This change is due to environmental and sustainability regulation in the last decades causes to considerable performance in green technology in the field of materials science through the development of bio

[3] To whom any correspondence should be addressed.

composites. There are several reasons for this interchanging to occur. Synthetic materials-based products pose a high risk in most applications, both from direct exposure and also from secondary exposure through soil, water and air pollution. In aspect of health, synthetic materials in fabrics are highly toxic and are increasing the negative effects on our health. These synthetic fabrics also pose a serious threat to ecological balance [1].

In order to overcome the problem, synthetic material needs to be replaced by natural fibres. Some examples of natural fibre that usually used in industries are rice husks, coconuts, kenaf, jute, oil palm and pineapple leaves. In this study, coconut coir and oil palm empty fruit bunch will be used as the natural fibres. The benefits of these fibres compared to fiberglass is, they are more environmentally friendly because it can be degraded naturally and even cheaper in price than fiberglass. Plus, glass fibre also gives bad effect to human health when the material is reuse because it will release carbon monoxide gas and dust to surrounding.

There are several methods used to determine types of composite compositions. One of them is functionally graded material (FGM) concept. FGM is a material which is its composition changed towards volume [2]. Functionally graded material (FGM) is a new and advanced material in this engineering generation. This is because of its speciality in its gradient composition and has a different structure from the original materials. Its manufacturing process involves a mixture of several materials to form a new composite or to change the strength of the materials [3]. This concept has a potential to be used in many sectors such as engineering, nuclear energy and electronic [4].

There are several techniques in producing FGM such as powder metallurgy, vapour deposition, rapid prototyping and centrifugal technique [2]. In this research, only centrifugal technique will be used in FGM producing because this is the only techniques that can form a continuous structure whether in radial or altitude specimen [2]. Centrifugal casting method has a several advantages such as have a good behaviour, quality and dimensional flexibility. This is because centrifugal casting method produces smaller microstructure than static casting method. At the same time, centrifugal reaction can eliminate the foreign materials from the molten and can reduce the dimension of molten shrinkage. The advantages and disadvantages of these approaches in producing FGNF are still investigated [5 - 7]. In this work, the preliminary study to characterize FGNF/epoxy fabricated using centrifugal casting is investigated.

2. Methodology

2.1 Specimen Preparation

Firstly, all the raw natural fibres were treated with water for 7 days and NaOH solution for 24 hours at room temperature. Then, those fibres were rinsed and dried under sunlight for 24 hours. Drying process continued using high capacity oven for 24 hours at 50°C. After that, all the fibres were cut into small pieces between 5-10mm lengths using granulator machine. The fibres then were mixed with epoxy and hardener before fabricated using centrifugal casting method. Specimens were divided into four compositions of natural fibres which are 0%, 5%, 10% and 15%. After all the materials were stirred, the mixture was poured into a rotating mould before it was put on laboratory roll ball mill. The mould rotated for 3 hours at 600 rpm speed. When the specimens were fully hardened, they were tested in several methods to characterize their microstructure view and mechanical behaviours. The FGNF/epoxy fabricated shown in figure 1.

Figure 1. FGNF/Epoxy fabricated with different contents of fibres (left to right): 0, 5, 10 and 15%.

2.2 Density and porosity test

The density determined by exploiting the principle of Archimedes, which states that anybody immersed in a fluid becomes lighter by an amount equal to weight of the fluid that has displaced. Based on ASTM D792 standard, the density, *D* and porosity, *P* were calculated by (1) and (2), respectively. *Ws*, *Wd* and *Ww* are denotes mass of an immersed and suspended specimen in water, mass of air-dried specimen, and mass of wet specimen, respectively.

$$D = \frac{Wd}{Wd - Ws} \quad (1)$$

$$P = \frac{Ww - Wd}{Ww - Ws} \text{x}100\% \quad (2)$$

2.3 Microstructural observation

Microscopic examination was conducted to study the structure of the material under magnification. The properties of materials determine how well they will perform under a given application, and these properties are dependent on the material's structure. Scanning Electron Microscopy (SEM) JEOL KSM-6380LA and Optical Microscopy (OM) were used to identify the failure fracture morphology of the composite specimen. Optical microscopy is an ideal method for general inspection purposes but scanning electron microscopy can provide the user with incredibly detailed topographical and compositional information.

2.4 Hardness test

Hardness test was run at room temperature according to ASTM D2240 standard. The tool that had been used in this test was shore durometer type D. This tool is usually used to test the hardness of polymer materials. Type D durometer was chosen to be used in this research because its function is to test a hard polymer material. The needle of the durometer will be pushed into the surfaces of the specimen radially starting from the inner, middle and outer part of the composite.

While, in longitudinal direction of the specimen were not been measured because it will have a uniform value of hardness along its height as it has a uniform distribution of fibres. Hardness value was recorded in one-minute reading. The specimen dimension needs to be 50mm height, 50mm width and 6mm thickness or more.

2.5 Compression test

The specimens were test at room temperature using Testometric Universal Testing Machine (UTM), modelled GOTECH 50KN. Force was set to 50 kN based on ASTM D695. The size of the specimen is 80 mm outer diameter, 5 mm thickness and 40 mm height. Actually, the size of the specimen was not following the testing standard size. However, the standard area ratio of the specimen (l/d) and the specification of the testing standard were still same which is 0.5. Before the testing start, the bottom surfaces of the specimen were mechanically planar grinded. This is to prevent the friction between the specimen and the surface of the testing machine plate while compressing process is running. Vertical force was given to the specimen with uniform speed, 1 mm/min until the specimen failed.

3. Results and discussions

3.1 Density and porosity

Table 1 tabulates the density and porosity of FGNF/epoxy fabricated. The lowest density recorded was at FGNF/epoxy composite specimen with 15% fibre content which was 1.1656 g/cm^3. The zero fibres content which was fully epoxy resin specimen was the highest density recorded followed by 5% and 10% fibre content. It can be concluded that increasing in fibre percentages will decrease the density of FGNF/epoxy composites.

Table 1. Density and porosity of FGNF/Epoxy fabricated.

Specimen	Density (g/cm^3)	Porosity (%)
0%	1.1782	12.73
5%	1.1769	13.5
10%	1.1738	14.34
15%	1.1656	15.32

Porosity measurement is important because it will show the critical imperfection of composite reinforced fibres. A higher percentage of porosity means many air gap or weak bonding between epoxy and fibres in the FGNF/epoxy fabricated. From table 1 can be seen that the increasing of percentage of natural fibres inside the specimens will caused the increasing of the porosity percentages. It is clearly showing that the existing of air gap or the separation between epoxy and fibres during centrifugal casting due to its different density where the higher density, epoxy, forced move to the mould outer periphery.

3.2 Microstructural observation

The microstructural observation is conducted to examine the fibres distribution and the bonding of fibres and epoxy matrix. It will determine the defect of these composites based on its fibres distribution through all the specimen, interface between matrix and fibres and its porosity.

Figures 2(a) show that the distribution of fibres (FGNF/epoxy with 5% fibre content) in all surfaces was balance. This was happened as the volume of the fibres was low and still can be flown by the epoxy resin smoothly while casting was running. Besides that, it shows that the middle and inner surfaces had the largest amount of porosity than outer surface of the specimen. This happened as the inner surface was exposed more to the air while casting than the outer surfaces, which was bounded by the inner mould surface.

Meanwhile for 10% fibre content, as shown in Figure 2(b), shows the moderate percentage of fibres. The figures show the appearance of fibres inside the FGNF/epoxy fabricated getting larger as the percentage was increased. This parameter also shows that the middle and the inner parts of the FGNF/epoxy fabricated have the largest amount of porosity. Porosity also might be happened while mixing and stirring activities were being performed.

Figure 2(c) shows the microstructure of the highest percentage of fibres content which was 15%. It shows the largest distribution of fibres inside the FGNF/epoxy fabricated especially at inner part of the specimen followed by middle and outer part. The inner and middle also had bigger size of pores then the outer part. From physical observation, there were few fibres accumulated at some part of FGNF/epoxy fabricated. This caused the weak bonding between matrix and fibres.

(a)
(b)
(c)

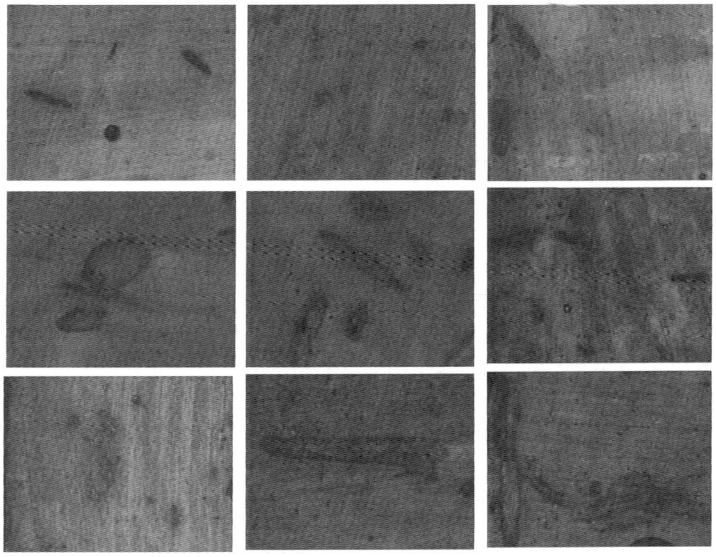

outer middle inner

Figure 2. Surfaces microstructure view of FGNF composite at various fibres content: (a) 5%, (b) 10% and (c) 15%.

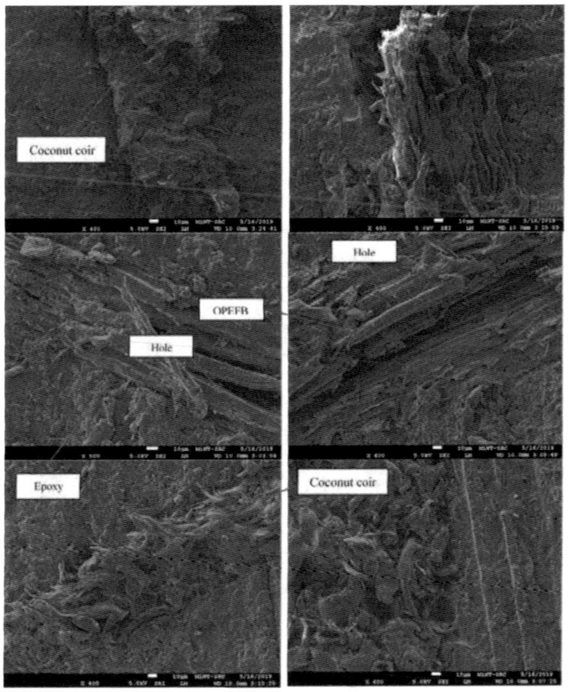

Figure 3. Microstructure view of samples from scanning electron microscope

Figure 3 shows that the microstructure views of samples using scanning electron microscope (SEM). It shows that every fibre inside the samples attached to the epoxy perfectly. This can be said because there was no wide gap between fibres and epoxy were displayed. Besides, there were no impurities presented on the fibres. This is because natural fibres were treated with water and NaOH solution earlier.

3.3 Hardness value

Based on hardness value (HD) of FGNF/epoxy fabricated measured along radial direction as shown in figure 4, it shows that the FGNF/epoxy with 5% fibres content has the highest value of hardness followed by 10%, 15% and 0%. The outer parts of all FGNF/epoxy fabricated are the hardest followed by the middle and inner part. The hardest value obtained is at outer part of 5% fibres content which is 84HD. While the lowest hardness value is at inner part of 0% fibres content which is 66.25HD. It shows that fibres additions in matrix enhance the hardness value especially to the middle part of the specimen as shown in figure 4. Besides, the centrifugal force introduced during casting play an important role in distributing fibres along radial direction that contribute to the variation of hardness value along radial direction of FGNF/epoxy fabricated. The FGNF/epoxy composite with 5% fibres content was the most balanced mixture to have a good hardness composite.

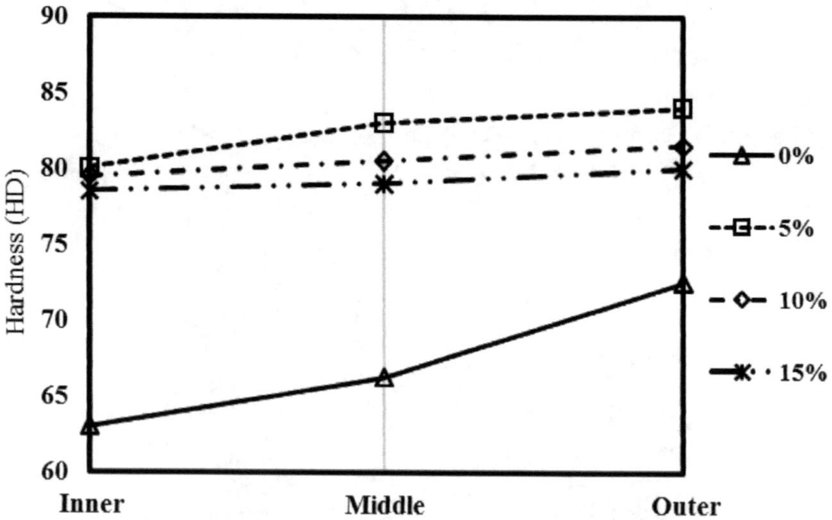

Figure 4. Hardness average (HD) versus specimen part graph

3.4 Compression strength

Figure 5 plot the load-displacement curves of the FGNF/epoxy fabricated with different percentages of fibres tested. No fibres addition specimen which was 0% fibres content also been tested as baseline. Compression force graph of all FGNF/epoxy fabricated start by ascending toward displacement until it reaches the ultimate strength of the specimens and then drop until its failure. For FGNF/epoxy with 5% fibres content, its graph drops drastically after reach the ultimate strength. While the graph for FGNF/epoxy with 0%, 10% and 15% fibres content decrease gradually toward displacement. This shows that FGNF/epoxy with 5% fibres content more brittle. While, the other specimens occurred strength loss until they failed this which is called plastic region.

Based on Table 2, it shows that the strength of FGNF/epoxy shows an increasing value from 0% to 5% fibres content, but it starts decreasing when to percentage of fibres content more than 5%. The decreasing in strength happened because of accumulation of fibres inside the specimen that blocked the epoxy matrix from mixed with the fibres. The data shows that specimen with 5% fibres content had the highest value of maximum stress which is 33.631 MPa. While, the lowest value of maximum stress was at 15% fibres content obtained which is 5.246 MPa. It can be concluded that FGNF/epoxy with 5% fibres content is the strongest composition in compression than others.

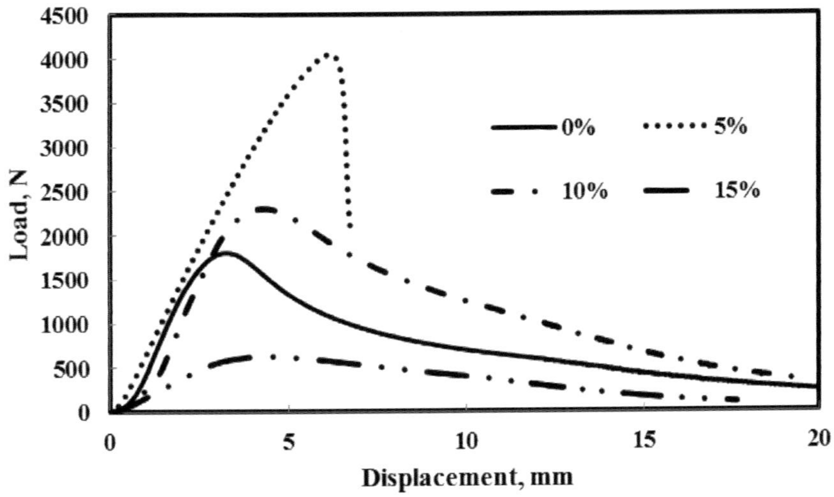

Figure 5. Load-displacement curve.

Table 2. Compression test data.

FGNF/epoxy	Max. Load (kN)	Max. Stress (MPa)
0%	17.690	15.016
5%	39.621	33.631
10%	22.577	19.164
15%	6.180	5.246

4. Conclusion

The cylindrical tube of functionally graded natural fibre reinforced polymer composite using centrifugal casting that have a better in strength was successfully fabricated. There were few differences between microstructure and mechanical behaviour of the composites as expected from the beginning of this research. The hardness and strength of FGNF/epoxy were increased with additional of natural fibres content from 0% to 5% but it was slightly decreased with additional of fibres content more than 5%.

However, the porosity is the main issue need to be reduced to enhance the composite. This may be contribute by the size of fibres. The smaller size of fibres may lead the good bonding of epoxy and fibres. This is happened when the size of the fibres is smaller, it will easily cover all the pores area inside the specimens. Thereby, porosity of samples can be decreased.

It can be determined that OPEFB and coconut coir are suitable to be used as polymer composite materials. It is one of new alternatives to replace synthetic fibre because they are environmentally friendly, not harmful to human, cheaper and easy to obtain.

Acknowledgement
Authors wishing to acknowledge financial support from Universiti Tun Hussein Onn Malaysia under TIER grant with vot no. U870.

References

[1] Singh Z and Bhalla S 2017 Toxicity of Synthetic Fibres & Health, *Advance Research in Textile Engineering* 2(1) 1–4
[2] Jamian S, Watanabe Y and Sato H 2019 Formation of compositional gradient in Al/SiC FGMs fabricated under huge centrifugal forces using solid-particle and mixed-powder methods, *Ceramics International* 45(7), 9444-9453
[3] Aboudi J, Pindera M J, and Arnold S M 1999 Higher-order theory for functionally graded materials, *Compos. Part B Eng.* 30(8) 777–832
[4] Tarlochan F 2012 Functionally Graded Material: A New Breed of Engineered Material, *J Appl Mech Eng* 1(5) 1–2
[5] Mahli M K, Jamian S, Ismail A E, Muhd Nor N H, Mohd Nor M K, Kamarudin K A and Awang M K 2018, Hardness and wear rate of Al LM6 hollow cylinder fabricated using horizontal centrifugal casting, *Journal of Physics: Conference Series* 1049 012042 1-7
[6] Jamian S, Ayob S N, Zainal Abidin M R and Muhd Nor N H 2016 Fabrication Of Functionally Graded Natural Fibre/Epoxy Cylinder Using Centrifugal Casting Method, *ARPN Journal of Engineering and Applied Sciences* 11(4) 2327-2331
[7] Jamian S, Razali S N F and Zainal Abidin M R 2016 FGNF/Epoxy Composite Fabricated Using Centrifugal Slurry-Pouring Method, *ARPN Journal of Engineering and Applied Sciences* 11(12) 7759-7764

IOP Conf. Series: Materials Science and Engineering **824** (2020) 012020 doi:10.1088/1757-899X/824/1/012020

Bottleneck-Based Heuristic for Permutation Flowshop Scheduling

N.A. Isa[a*], S.A. Bareduan[b] and A.S. Zainudin[c]

[1] Faculty of Mechanical and Manufacturing Engineering,
Universiti Tun Hussein Onn Malaysia,
86400 Parit Raja, Batu Pahat, Johor, Malaysia

E-mail: [*a]nooramira1992@gmail.com, [b]saleh@uthm.edu.my,
[c]amirasyu27@gmail.com

Abstract. This paper addresses the permutation flowshop scheduling problem with objective of makespan minimization. It proposes a heuristic that apply bottleneck-based identification technique and Nawaz, Enscore and Ham (NEH) insertion technique. This experiment was carried out using measurement of four-machines and six-jobs simulated in Microsoft Excel Simple Programming. The computational result show that bottleneck-based (BNB) achieved 60 % and more of optimum solution and heuristic validation result prove that the proposed heuristic resulting to be much better than existing heuristic by achieving more best solution of 6 sets out of 10 sets replication.

1. Introduction

Scheduling is an act of planning and prioritizing activities with the time limit for the job completion in order to meet with certain requirement, constraints or to achieve the goal of objectives [1]. Since the time always be the greatest constraint, it is why the scheduling is important to organize the activities efficiently and optimally. In worldwide industries, scheduling plays an important role in the production system since the resources are becoming more critical to be controlled. The resources are referred to the machine, manpower, material and many more. The greatest outcome can be gained if the scheduling of the resources is successfully organized [2]. It will be a great advantage to the manufacturer in worldwide industries if the latest research were successful to perform, easy to understand and easily applied to the current manufacturing system.

Permutation flowshop scheduling problem (PFSP) is one of the best known production scheduling problems with the same job order on all machines which has an intense engineering background [18]. Johnson [4] becomes the pioneer by introducing the Johnson's rule with two machine problem. However, the general problem has been proved to be strongly NP-hard when involving more than two machine problem, $m > 2$ [4-8]. Minimization of makespan becomes one of the most attractive objective where it leads to the development of many approximate algorithms. NEH heuristic appeared around three decades ago are known as the highest performing method in minimizing the makespan for permutation flowshop scheduling problem. NEH algorithm was proposed by Nawaz, Enscore and Ham [5-7], [17]. This algorithm gave a highest attention on the job with larger total processing time where it should have higher priority. The jobs are arranged in decreasing order of the total processing time and

IOP Conf. Series: Materials Science and Engineering **824** (2020) 012020 doi:10.1088/1757-899X/824/1/012020

the 3rd job in the job list is inserted into the previous partial sequence at the position that gives the best mean flow time or makespan. Many researchers have used NEH heuristic for comparison purpose since it had appeared to be the best heuristic if the makespan criterion is considered [6-7]. Some of them also modified the NEH heuristic following to their objective function, [5], [12], [15-17], [19], [21-22].

In the scheduling environment, shifting bottleneck model has been widely used in sorting the machine with the highest increase of objective function value [9]. According to Zhenqiang and Peng [10], identification of bottleneck resources helps in scheduling the jobs rationally in reducing the difficulty in follow-up scheduling and also ensures the feasibility and effectiveness of scheduling result. Basically, the main idea is to schedule the jobs at the bottleneck stage where it may affect the performance of a heuristic for scheduling jobs in all the stages [11]. In the study of Modrak and Pandian [12] bottleneck machines were detected based on the machine workload, utilization rate or idle time length which bottleneck involvement significantly affect the quality of final solution. Utilization bottleneck heuristic approach is known to be among the most successful methods in solving shop scheduling problem including the shifting bottleneck heuristic and bottleneck minimal idleness heuristic [13].

In this paper, we investigated the bottleneck based concept in finding a good job arrangement for flowshop scheduling problem with the objective of makespan minimization. This study performed the concept by using a small problem size of four-machines and six-jobs and was carried out by using Microsoft Excel simple programming in Visual Basic for Application (VBA). This concept was applied on 100 sets of randomly generated data in 10 replications. The effectiveness of this concept was evaluated against the optimum solution based on the average percentage error, standard deviation and the number of best solution obtained [4], [15], [20-21], [23]. This proposed heuristic was also compared with existing heuristic, NEH heuristic.

2. Methodology

The bottleneck-based heuristic firstly identifies the bottleneck stage and then determines the schedule of the jobs in the bottleneck stage [11]. The job processing times are used as indicator to determine the bottleneck stage. The procedures of the BNB heuristic are presented as below:

2.1 Bottleneck Machine Identification Procedure

Based on the four machines problem, there will be four possible dominant machines and each of the dominance will have their own initial sequences. The initial sequence for each of the machine dominance was arranged based on the result of machine dominant identification. Bottleneck identification is important to classify the criticality of process machines. The machine criticality or called as dominant machine is classified based on the processing times of job being process in each machines. 100 sets of processing times are generated randomly in VBA. All the steps to identify the machine bottleneck are explained below:

Step 1: The average processing time for all the six jobs on four machines was calculated as shown in Figure 1.

PROCESSING TIME DATA(HOURS)					
JOB	M1	M2	M3	M4	TOTAL
A	13	28	11	36	88
B	23	30	34	6	93
C	11	28	7	32	78
D	22	15	35	37	109
E	34	34	28	39	135
F	10	5	28	27	70

AVG Process	25.150

Figure 1. Example of processing time data.

Step 2: The domination of machine was then determined using following equations:

$$X = \begin{cases} 1, & \text{if } t > t_{avg} \\ 0, & \text{if } t \leq t_{avg} \end{cases} \tag{1.0}$$

Where;

X= Dominance value

t = Processing time

t_{avg}= Average processing time

Step 3: The total dominance value was calculated for each machine as shown in Figure 2. The machine which has the highest dominance value is identified as the bottleneck machine.

	Dominance calculation			
	M1	M2	M3	M4
	0	1	0	1
	0	1	1	0
	0	1	0	1
	0	0	1	1
	1	1	1	1
	0	0	1	1
Total Dominance	1	4	4	5

Max.Dominance value	5
Dominance machine	M4

Figure 2. Example of dominance calculation.

2.2 Initial Sequences and Insertion Phase

In the study of Framinan and Rajendran [7], it was found that the initial job arrangement and the opportunity of inserting a job in any possible position are the strength of NEH heuristic which known as the best heuristic in solving the permutation flowshop problem with objective of makespan minimization. Thus, in this paper, the initial sequence was arranged in best arrangement based on the result of machine dominant identification so that the final solution will have the best answer.

- In cases of machine 1 dominant (M1), the processing time of jobs for machine 1 was used (M1).
- In cases of machine 2 dominant (M2), the processing time of jobs for machines 1 and 2 was sum up (M1+M2).
- In cases of machine 3 dominant (M3), the processing time of jobs for machines 3 and 4 was sum up (M3+M4).
- In cases of machine 4 dominant (M4), the processing time of jobs for machines 4 was used (M4).

From the observation, in building an initial sequence for all dominant machine, it was found that for M1 dominant, the machine criticality is depending on the processing time at machine 1. For M2 dominance, the machine criticalities are on the machine 1 and 2, thus the processing time for both machines was summed up. Next, for M3 dominance, the processing time for the second last and the last machine was considered while for M4 dominance, the processing time of last machine is the most critical. The sequence guide is the job with highest processing time must be processed first and then proceed with the job with second highest processing time and continue with the next job until the last job with lowest processing time. A job with highest processing time must be processed first so that there are not much idle time towards the end of the overall sequence.

The generalization sequences of this heuristic can be described as follows:

Input:

Suppose a standard permutation flowshop problem with i = 1,..., m machines, j = 1,…, n jobs, processing time Pij and the objective function is makespan minimization, Cmax.

Step 1: Sort the *n* jobs according to descending sums of processing time depending on the machine domination cases.

Step 2: The initial sequences arrangement are depending on the job sorting in Step 1. The first place was filled by a job with largest sum of processing time, Max ΣPij followed by the second largest sum of processing time and so on until the last place, the sixth place was filled with the lowest sum of processing time, Min ΣPij.

Step 3: Then, take the first two jobs from the job arrangement in Step 2 and schedule their start-stop time. Reverse the position of jobs. Determine the first partial sequences with the lowest objective function, Cmax.

Step 4: Next, for the *k*th job, $k = 3,.., n$, insert the job into the place, among the position 1, 2,…, k job of partial sequences, while keeping the relative sequence of partial sequence. Choose the best sequences out of *k* sequences as partial sequence for the next iteration.

3. Results and Discussion

This section discussed the result of the BNB heuristic performance through comparison with the optimum answer and also the NEH heuristic. A test on heuristic validation was also explained in this section. This paper performed the experiment using the problem size of four-machines and six-jobs. There are 4 potential machines dominance in this paper and the processing time are randomly generated using uniform distribution for 100 set of test problems with 10 replications.

3.1 Heuristic Performance

Table 1 summarizes the factors used to define the experimental condition: number of machines, number of jobs, number of bottleneck machine, variation of job processing time, number of test problems and number of replications. At the stage of bottleneck identification, the result shows that the machine dominance 1, M1= 35 data set, machine dominance 2, M2= 19 data set, machine dominance 3, M3= 27 data set, machine dominance 4, M4= 19 data set. The result was shown in Table 2.

Table 1. Experimental design of generating test problems.

Experiment Condition	Levels
Number of machines	4
Number of jobs	6
Number of bottleneck machine	4
Variation of job processing time	[1,40] (low, high)
Number of test problems	100
Number of replications	10

Table 2. Machine Domination.

Machine Dominance	Number of data set
M1	35
M2	19
M3	27
M4	19

In this paper, in order to evaluate the effectiveness of this proposed bottleneck-based (BNB) heuristic, the simulated result was compared with the optimum answer based on the average percentage error (Avg), standard deviation (Std) and the number of optimum solution obtained (Best). The optimum makespan was recorded from the complete 720 iterations. The error for each heuristic solution is defined as 100[(Heuristic Best)/ Best]. The average error is then computed for each dominant group and this is summarized in Table 3. The evaluations of BNB heuristic have been made against NEH heuristic. The different of BNB against NEH heuristic is defined as 100[(BNB – NEH)/ NEH] and was classified based on the dominant group, the results was tabulated in Table 4.

Table 3. BNB heuristic evaluation against optimum answer.

Machine Dominance	Avg error (%)	Std	Best solution obtained	Optimum result (%)
M1	1.09	0.0196	21	60.00
M2	1.13	0.0194	12	63.16
M3	0.85	0.0190	21	77.78
M4	0.32	0.0062	13	68.42

Table 4. Evaluation of BNB heuristic against NEH heuristic.

Machine Dominance	BNB VS NEH			Average difference BNB/NEH (%)
	WIN	TIE	LOSE	
M1	11	20	4	-0.48
M2	3	13	3	-0.20
M3	4	22	1	-0.19
M4	5	13	1	-0.45

From the result Table 3, the average percentage errors and the standard deviation for BNB heuristic against the optimum makespan are quite small since the BNB completion times are very close to optimum result. It clearly showed that the number of best solutions obtained by BNB heuristic are impressive. Machine dominance 1 (M1) achieved 21 best solutions from 35 solutions, M2 achieved 12 best solutions from 19 solutions while M3 achieved 21 best solutions from 27 solutions and lastly M4 achieved 13 best solutions from 19 solutions. As expected, the optimum result showed that the proposed heuristic is performing well since it achieved 60% and more of optimum answer.

Table 4 shows the evaluation of performance of BNB heuristic against NEH heuristic. Since the BNB heuristic achieved good results against the optimum solution, it is expected that this heuristic can compete with existing heuristic, NEH heuristic. From the result of BNB heuristic against NEH heuristic, machine dominance M1 produced 11 better makespan (11 wins), 4 worse makespan (4 lose) and 20 equal makespan (20 tie) compared to the NEH. M2 produced 3 wins, 3 lose and 13 ties makespan result against the NEH. For M3 and M4, the results are as shown in Table 4. Overall, it was found that BNB heuristic have gained 67 number of optimum solutions compared to NEH heuristic which only gained 58 number of optimum solutions. The average different of BNB heuristic against NEH heuristic shows the negative value because the solution produce by BNB is better than the NEH solution for each of the machine dominance. So, the table clearly shows that BNB heuristic performs better than the NEH heuristic.

3.2 Heuristic Validation

In order to validate the heuristic performance, a test on 10 replications of 100 set of random data was carried out. The performance of both heuristic, BNB heuristic and NEH heuristic was measured based on the best solution obtained (Best) and also the average percentage error (Avg). The results are shown in Table 5.

Table 5. BNB heuristic against NEH heuristic performance.

Replication Set of Random Data	Best solution obtained		Average Error (%)	
	BNB	NEH	BNB VS OPT	NEH VS OPT
1	67	58	1.00	1.35
2	64	62	1.02	0.77
3	65	73	1.05	0.82
4	65	64	0.98	0.93
5	62	62	0.94	0.94
6	61	68	0.95	0.88
7	67	63	0.88	1.00
8	61	60	1.13	0.96
9	65	68	1.09	0.76
10	67	64	0.98	0.89

From the result in Table 5, the best solution obtained by the BNB heuristic is much better than the NEH heuristic. BNB heuristic achieved more best solution in 6 set of replication, loss best solution in 3 set of replication and tied in 1 set of replication. BNB heuristic produces an answer more than 60 best solution compared to the NEH. Even though the BNB heuristic obtained more best solution than NEH, however, from the overall result of average percentage error, the result shows that NEH heuristic perform a little better in producing a good solution close to optimal answer. From the observation, it was found that BNB heuristic still can be improved to produce an answer near to the optimal answer since the job arrangement are not effective to a certain sets of data where it resulted into large completion time.

4. Conclusion

This paper addresses the permutation flowshop scheduling problem with the objective of makespan minimization. Many researches and heuristics have been proposed to compete with the existing heuristic. This paper proposes a heuristic for a small size flowshop problem which the performance of the heuristic is depending on the quality of initial sequence. The problems set are first being classified into specific number of bottleneck-based machines. This paper applied the NEH insertion technique since NEH heuristic are the most famous heuristic in solving permutation problem objective function. From the results that have been shown in the previous section, bottleneck-based (BNB) achieved 60 % and more of optimum solution and heuristic validation result achieved more best solution of 6 sets out of 10 sets replication. Overall, it can be concluded that the proposed BNB heuristic performs better than the NEH heuristic.

Acknowledgement

This research is funded by Ministry of Higher Education of Malaysia (MOHE) and Universiti Tun Hussein Onn Malaysia (UTHM) under Fundamental Research Grant Scheme (FRGS, Vot 1497).

References

[1] Sule, D. R. 2008. *Production Planning and Industrial Scheduling; Examples, Case Studies and Apllications; Second Edition*. United States of America: CRC Press, Taylor and Francis Group.

[2] Thomas, E. M. & David, W. P 1993. *Heuristic Scheduling Systems.* Canada: Wiley-Inter science.

[3] Johnson, S. 1954. Optimal two-and three stage production schedules with setup times included. *Naval Research logistics Quarterly.*1:61.

[4] Allahverdi, A. and Aldowaisan, T. 2002. New heuristics to minimize total completion time in *m*-machine flowshops. *International Journal Production Economics.* 77: 71-83.

[5] Kalczynski, P. J. and Kamburowski, J. 2008. An improve NEH heuristic to minimize makespan in permutation flow shops. *Computers & Operations Research.* 35: 3001-3008.

[6] Kalczynski, P. J. and Kamburowski, J. 2007. On the NEH heuristic for minimizing the makespan in permutation flow shops. OMEGA, The International Journal of Management Science. 35: 53-60.

[7] Framinan, J. M., Leisten, R. & Rajendran C. 2003. Different intial sequences for the heuristic of Nawaz, Enscore and Ham to minimize makespan, idletime or flowtime in the static permutation flowshop sequencing problem. *International Journal of Production Research.* 41(1), pp. 121-148.

[8] Modrak V. and Pandian R. S. 2010. Flow shop scheduling algorithm to minimize completion time for *n*-jobs *m*-machine problem. *Technical Gazette.* 17(3), pp. 273-278.

[9] Braune, R and Zapfel,G. 2015. Shifting bottleneck scheduling for total weighted tardiness minimization- A computational evaluation of subproblem and re-optimization heuristics. *Computers & Operations Research.* 66: 130-140.

[10] Zhenqiang, B., Weiye, W., Peng, W., and Pan, Q. 2012. Research on production scheduling system with bottleneck based on multi-agent. *Physics Procedia.* 24: 1903-1909.

IOP Conf. Series: Materials Science and Engineering **824** (2020) 012020 doi:10.1088/1757-899X/824/1/012020

[11] Chen, C. L. and Chen C. L. 2009. A bottleneck-based heuristic for mimizing makespan in a flexible flow line with unrelated parallel machines. *Computers & Operations Research*. 36: 3073-3081.

[12] Zhang, R. and Wu, C. 2012. Bottleneck machine identification method based on constraint transformation for job shop scheduling with genetic algorithm. *Information Sciences*.188: 236-252.

[13] Bareduan, S. A. and Hasan, S. 2009. Bottleneck-Based Heuristic For Flow Shop With Two Potential Dominant Machines. *Computers & Industrial Engineering*. 978: 4136-4244.

[14] Allahverdi, A. and Aldowaisan, T. 2002. New heuristics to minimize total completion time in *m*-machine flowshops. *International Journal Production Economics.* 77: 71-83.

[15] Aldowaisan, T. and Allahverdi, A. 2003. New heuristics for no-wait flowshops to minimize makespan. *Computers & Operations Research*. 30: 1219-1231.

[16] Kalczynski, P. J. and Kamburowski, J. 2005. A heuristic for minimizing the makespan in no-idle permutation flow shops. *Computers & Industrial Engineering*. 49: 146-154.

[17] Viagas, V. F. and Framinan, J. M. 2015. NEH-based heuristic for the permutation flowshop scheduling problem to minimise total tardiness. *Computers & Operations Research*. 60: 27-36.

[18] Abedinnia, H. Glock, C. H. and Brill, A. 2016. New simple constructive heuristic algorithms for minimizing total flow-time in the permuatation flowshop scheduling problem.*Computers & Operations Research*. 74: 165-174.

[19] Framinan, J. M., Leisten, R. & Rajendran C. 2006. A heuristic for scheduling a permutation flowshop with makespan objective subject to maximum tardiness. 99: 28-40.

[20] Chakravarty, K. and Rajendran, C. 1999. A heuristic for scheduling in a flowshop with bicteria of makespan and maximum tardiness minimization. *Production Planning and Control*.10: 707-14.

[21] Aldowaisan, T. and Allahverdi, A. 2003. New heuristics for no-wait flowshops to minimize makespan. *Computers & Operations Research*. 30: 1219-1231.

[22] Allahverdi, A. and Al-Anzi, F.S. 2006. A branch-and-bound algorithm for three-machine flowshop scheduling problem to minimize total completion time with separate setup times. *European Journal of Operational Research*. 169: 767-780.

[23] Allahverdi, A. 2004. A new heuristic for m-machine flowshop scheduling problem with bicriteria of makespan and maximum tardiness. *Computers & Operations Research*. 31: 157-180.

IOP Conf. Series: Materials Science and Engineering **824** (2020) 012021 doi:10.1088/1757-899X/824/1/012021

Dynamic Weighted Idle Time Heuristic for Flowshop Scheduling

A.S. Zainudin[1,a] , S.A. Bareduan[1,b*] and N.A. Isa[1,c]

[1] Faculty of Mechanical and Manufacturing Engineering,
Universiti Tun Hussein Onn Malaysia,
86400 Parit Raja, Batu Pahat, Johor, Malaysia

E-mail: [a]amirasyu27@gmail.com, [*b]saleh@uthm.edu.my and [c]nooramira1992@gmail.com

Abstract. Flowshop scheduling is used to determine the optimal sequence of n jobs to be processed on m machines in the same order. The objective of this paper is to minimize the makespan for group of jobs waiting in the process system. It proposed a new heuristic that apply dynamic weighted idle time (DWIT) that used several weight factor values as a multiplication factor of the total idle time. The result shows that the average percentage errors and standard deviation for DWIT heuristic are smaller compared to the NEH (Nawaz, Enscore and Ham) and it achieved more than 50% of optimum answer. Therefore, this DWIT heuristic has great potential to outperform the NEH heuristic.

1. Introduction

Scheduling is one of the important areas in the field of production management. Scheduling flowshop problem can be addressed as the setting with penalties for tardiness in delivering customer orders, as well as cost for holding both finished goods and work-in-process inventory [1,2]. In a typical real-world scheduling problem, the set of jobs changes dramatically over time and the processing times of jobs are affected by various types of uncertainty. The goal is to determine how the available machine processing time is to be allocated among competing requests with the objective of optimizing the performance of the system. In general, methods for solving dynamic scheduling problems must address the combinatorial structure inherited in most interested scheduling problems [3].

Efficient scheduling is how manufacturing companies minimize the cost and punctuality to meet customer with the promised due date [4]. Although there has been an increasing interest in modelling and solving scheduling problems in dynamic and uncertain environments, scheduling research has mostly focused on devising effective method for solving deterministic problems with a complex combinatorial structure [5-7]. On the other hand, scheduling problems with a simpler combinatorial structure but with stochastic and dynamic characteristic have been studied for a long time.

The scheduling objective is to sequence the processing of waiting jobs on the machines so that some overall performance measure is optimized [8].This paper analyses scheduling heuristic identified as dynamic weighted idle time heuristic in flowshop environments. The objective in this case is to minimize the completion time or the makespan for group of jobs waiting in the process system.

IOP Conf. Series: Materials Science and Engineering **824** (2020) 012021 doi:10.1088/1757-899X/824/1/012021

2. Flowshop scheduling problem

Job scheduling is an important tool in production management. It is useful in increasing the production of the product, meet the demands of the market in time and to minimize the flow time and cost of the product [9]. The PFSP was first studied by Johnson in 1954 and since then, many heuristic and metaheuristic methods for its solution have been proposed [10]. Flow shop scheduling deals with a minimal downtime and minimal waiting time. It is special case of the shop scheduling tasks where there is a strict order of all operations performed on all jobs [11].

Flowshop scheduling problem can be classified into two categories namely with and without an operation interval waiting time. In a flowshop system with waiting times, the jobs are processed from one machine to the next one allowing waiting time in between, whereas in a no wait flowshop scheduling, the jobs will go for the processing from one machine to the next machine without waiting time. Therefore, in classical flowshop sequencing problem with waiting time jobs may be queued in front of each machine. In such case, an unlimited buffer is considered at the front of each machine [12].

Flowshop scheduling is a production system that consists of *m* number of machines or processors arranged in series. Jobs that are processed in a flowshop follow a common route through the shop. It begins with the first machine and then ending at m^{th} machine, visiting each of the machines only once and in the same order. Although all jobs follow the same route, the processing time requirements for their operations on each of the machines may differ. After completion of a job's operations on one machine, the job is then transferred to the next machine in sequence. If the machine is occupied with another job, the transferred job is deposited in queue at a buffer to wait for processing. It is assumed that buffer space is always available to hold any number of waiting jobs at any machine in the flowshop scheduling [8,13].

Generally, in a scheduling it is supposed to have *m* machines to process *n* jobs. A schedule assigns for each job an allocation of one or more time intervals to one or more machines [14]. Schedules also may be represented by Gantt charts [15]. Most of researcher has focused on the flow shop scheduling which can be considered as a classical flow shop problem by adding the assumption that the jobs must be processed in the same sequence by each of the *m* machines [14].

This flowshop scheduling problem can be described as follows. Each job j_i (i = 1,2,…,n) has to be processed on m machines M_j (j = 1,2,…,m) following the same order in all machines. The processing time of job j_i on machine M_j is known as p_{ij}. In solving the problem, the major considered constraints are as the following:

- All jobs are ready for processing at time zero.
- The machines are continuously available from time zero onwards (no breakdowns).
- At any time, each machine can process at most one job and each job can be processed on at most one machine.
- No pre-emption is allowed (once the processing of a job on a machine has started, it must be completed without interruption).

3. Example of flowshop scheduling problem

In order to illustrate the flow shop scheduling problem, let's consider the following example. Let's there be six jobs of which to be processed at four machines flowshop. The workflow diagram of this problem is shown in Figure 1.

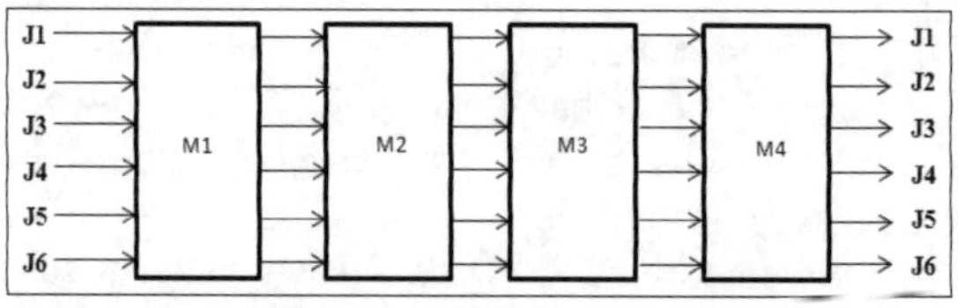

Figure 1. Workflow in flow shop scheduling.

IOP Conf. Series: Materials Science and Engineering **824** (2020) 012021 doi:10.1088/1757-899X/824/1/012021

The example of processing time matrix (in hours) is given in Table 1. For the six jobs and four machines, the total possible schedules of the jobs are 720 solutions (6!). Figure 2 shows the gantt chart that represent the example of the flowshop scheduling with the sequence B,E,D,A,F,C in decreasing arrangement according to the total processing time in Table 1.

Table 1. Processing time matrix.

	Processing Time Data (Hours)				
Job	Machine 1	Machine 2	Machine 3	Machine 4	**Total**
A	22	12	24	21	**79**
B	36	15	11	40	**102**
C	9	7	30	11	**57**
D	24	33	23	20	**100**
E	38	38	5	20	**101**
F	19	5	23	14	**61**

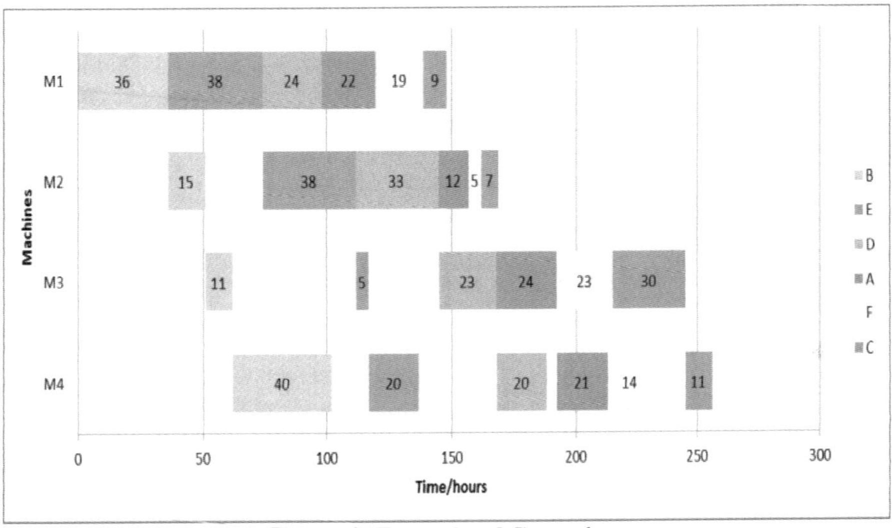

Figure 2. Example of Gantt chart.

4. Dynamic weighted idle time heuristic

Several weight factor values have been assigned and were used as a multiplication factor of the total idle time for machine 2, machine 3 and machine 4. Weight factors have been created based on the example of gantt chart in Figure 2. This study only considered the idle time at machine 2 until machine 4 with exception on machine 1 because there is no idle time occurred at machine 1. The range of weight factor value is from 0 to 1. Three levels of weight factor are considered in this study. First level is zero, second is low level and high level. At machine 2, zero level was used because the idle time at this machine is anticipated to give the least impact to the overall schedule makespan. Next, machine 3 has been placed with a low level that is within 0 until 0.4 and then a high level with the range 0.8 until 1.0 was used in machine 4 because it is anticipated the idle time of the last machine will have high impact to the overall scheduling makespan. Table 2 shows the example of weight factor value.

Table 2. Example of weight factor value.

Machine	Weight Factor Value				Level
	First Phase	Second Phase	Third Phase	Fourth Phase	
Machine 2	0	0	0	0	Zero
Machine 3	0	0.1	0.3	0.1	Low
Machine 4	1.0	0.8	0.9	1.0	High

The weight factors were applied with trial and error method until the objective of this study to minimize the makespan is achieved. Four sets of weight factors was assigned and become a multiplication factor for total idle time from the first two jobs until the fifth job sequence step. From the Table 2 above, the weight factor values for the first phase were assigned for the first two jobs sequencing while second phase was for third job sequencing until the fourth phase for fifth job sequencing. This study only used four sets of weight factor because for the last job the minimum makespan was chosen as the sequence selection criteria instead of idle time.

The implementation steps for this heuristic are as follows:
Step 1: Sort the n jobs in non-decreasing order of the total processing times.
Step 2: Take the two earliest jobs from the job arrangement in step 1 and schedule their start-stop time. Then, reverse the position of jobs. The consideration of idle time was idle time before and in-between two jobs only. The idle time after completion of the last job was not involved.
Step 3: The sum of idle time for each machine was multiplied with the weight factor. The minimum idle time after multiplication of weight factor was chosen as partial sequence for the next sequencing step.
Step 4: For the next job consider k=3,…,n, insert the non-scheduled job by placing the job with all the possible positions in the previous sequence. The relative positions of the jobs that already assigned shall not be changed. Choose the best sequences out of k sequence for the next iteration.
Step 5: At the sixth job, the best sequence was chosen based on the minimum makespan.

5. Results and discussion

In order to conduct an evaluation on the ability of the weighted idle time method in solving the flowshop scheduling problem, a test for a random data was necessary. This study used 100 sets of randomly generated data with 10 replications for the evaluation process. In order to obtain the optimum makespan solution, a simple programming was developed in Microsoft Excel to run all the 720 iterations of all possible solutions for each set of data by using Visual Basic Application programming. Table 3 shows the example of 100 sets of randomly generated data. The table indicate optimum (OPT) makespan, NEH makespan, Dynamic Weighted Idle Time (DWIT) makespan and percentage error between NEH against Optimum and DWIT against Optimum. The percentage error are defined as [(Heuristic-OPT)/OPT]x100.

Table 3. Performance of DWIT heuristic against NEH heuristic.

Data	Makespan OPT	Makespan NEH	Makespan DWIT	Error NEH VS OPT (%)	Error DWIT VS OPT (%)
1	204	204	204	0.00	0.00
2	199	207	204	4.02	2.51
3	216	220	220	1.85	1.85
4	214	214	214	0.00	0.00
5	199	203	203	2.01	2.01

6	186	186	186	0.00	0.00
7	226	226	226	0.00	0.00
8	192	192	192	0.00	0.00
9	194	194	194	0.00	0.00
10	195	201	201	3.08	3.08
-					
-					
-					
99	175	182	182	4.00	4.00
100	147	151	151	2.72	2.72

Table 4 shows the summarization result that has obtained from the generated data. The results displayed the average percentage error of NEH and DWIT. Average percentage error of NEH was 1.24% and DWIT was 1.10% while the standard deviation of NEH was 0.019 and DWIT was 0.017. From the result, its shows that average percentage errors and the standard deviation for DWIT heuristic has little error compared to the NEH since DWIT gained 61 optimum answers compared to the NEH which only gained 58 optimum answers. Therefore, this result point that the proposed heuristic is performing well since it achieved more than 50% of optimum answer.

From the performance shows in Table 3 and Table 4, DWIT heuristic has an ability to compete with NEH heuristic. Therefore, this study has carried out 10 replications of 100 set of random data for the validation process as shown in Table 5. It represent the average error of NEH and DWIT, standard deviation (Std) of NEH and DWIT, the best solution which defined as makespan heuristic is equal to the optimum answers and also the summarization result of DWIT against NEH.

Table 4. Summarization result DWIT heuristic against NEH heuristic.

Average Percentage Error		**Standard Deviation (Std)**		**Best solution obtained**	
NEH	**DWIT**	**NEH**	**DWIT**	**NEH**	**DWIT**
1.24%	1.10%	0.019	0.017	58	61

Table 5. Heuristic validation result.

Replication Set of Random Data	**Average Error NEH (%)**	**Average Error DWIT(%)**	**Std of NEH**	**Std of DWIT**	**Best solution obtained**		**DWIT against NEH**
					NEH	**DWIT**	
1	1.24	1.01	0.019	0.017	58	61	Better
2	0.82	0.91	0.018	0.019	73	71	Worse
3	1.00	1.01	0.017	0.016	63	62	Worse
4	0.88	0.92	0.018	0.018	68	69	Better
5	0.85	0.85	0.013	0.013	60	60	Tie
6	1.05	1.01	0.018	0.017	61	64	Better
7	1.12	1.15	0.018	0.012	60	61	Better
8	0.78	0.86	0.016	0.016	65	61	Worse
9	0.94	0.96	0.016	0.017	62	62	Tie
10	1.12	0.89	0.019	0.019	67	69	Better

In Table 5, it shows that there are five replications that have better result, two replications tie and three replications worse when compared with NEH heuristic. From the five replications better result, only three replications have smaller average error than the NEH. Besides, for the standard deviation, from the 10 replications there are 4 replications data that have lower standard deviation than NEH heuristic, while 4 replications have same standard deviation with NEH and the other has higher standard deviation than NEH. However, the different of standard deviation between DWIT and NEH heuristic are not much differs. The result shows that the proposed heuristic has a great potential to compete with the NEH.

6. Conclusion

This paper employs the dynamic weight factor in order to get the minimum makespan in flowshop scheduling. A new algorithm was developed to solve this flowshop problem. In general, the preliminary results revealed that the proposed heuristic give an optimal sequence which minimize the makespan of the problem. This DWIT heuristic has a great potential to outperform with the NEH heuristic if the improvement of dynamic weighted idle time is conducted with a better weight factors. Further study is being carried out with different weight factors with more testing data.

Acknowledgement

This research is funded by Ministry of higher education of Malaysia (MOHE) and Universiti Tun Hussein Onn Malaysia (UTHM) under Fundamental Research Grant Scheme (FRGS, Vot 1497).

References

[1] Bulbul, K., Kaminsky, P. & Yano, C. Flowshop Scheduling With Earliness, Tardiness and Intermediate Inventory Holding Costs. Naval Research Logistics. 2004;51

[2] Kemppainen, K. Priority Scheduling Revisited-Dominant Rules, Open Protocols, and Integrated Order Management. 2005.

[3] Terekhov, D., Tran, T. T. & Beek, J. C. Investigating Two-Machine Dynamic Flow Shops Based on Queuing and Scheduling. 2013.

[4] Heizer, J. & Render, B. Operations Management: Sustainability and Supply Chain Management. Prentice Hall. 2014.

[5] Bidot, J., Vidal, T., Laborie, P. & Beck, J. C. A Theoretic and Practical Framework for Scheduling in a Stochastic Environment. Journal of Scheduling 2009;12(3):315-344.

[6] Aytug, H., Lawley, M. A., Mckay, K., Mohan, S. & Uzsoy, R. Executing Production Schedules In The Face Of Uncertainties: A Review and Some Future Directions. European Journal of Operational Research. 2005;161:86-110.

[7] Suresh, V. & Chaudhuri, D. Dynamic Scheduling- A Survey of Research. International Journal of Production Economics. 1993;32:53-63.

[8] El-Bouri, A. The Effect of Inserted Idle Time on the Performance of Dispatching Rules In A Flowshop. Proceedings of the 2014 International Conference on Industrial Engineering and Operations Management Bali, Indonesia. 2014.

[9] Gupta, S. Weighted Mean Flow Time In No-Wait Uncertain Scheduling Under Various Parameters. International Journal of Computational Intelligence Techniques. 2012;3:99-101.

[10] Ruiz, R. & Stutzle, T. A Simple And Effective Iterated Greedy Algorithm For The Permutation Flowshop Scheduling Problem. European Journal Of Operational Research. 2007:2033-2049.

[11] Chandradansena, S., Dubey, D. & Khandelwal, A. K. Reduce the Time Of Multi Job Operation By Flowshop And Gantt Chart. International Research Journal of Engineering and Technology (IRJET). 2005;2.

[12] Grabowski, J. & Pempera, J. The Permutation Flow Shop with Buffers: A Tabu Search Approach. European Journal of Operational Research. 2000;116:205-219.

[13] Tseng, C. T. & Liao, C. J. A Discrete Particle Swarm Optimization for Lot-Steaming Flowshop Scheduling Problem. European Journal of Operational Research. 2007;191:360-373.
[14] Kouki, S., Jemni, M. & Ladhari, T. Solving The Permutation Flow Shop Problem With Makespan Criterion Using Grids. International Journal of Grid And Distributed Computing. 2011;4.
[15] Brucker, P. Scheduling Algorithm, New York, Springer. 1998

AUTHOR INDEX